Frictional Electricity

by W. Jerome Harrison

with an introduction by Roger Chambers

This work contains material that was originally published in 1895.

This publication was created and published for the public benefit, utilizing public funding and is within the Public Domain.

This edition is reprinted for educational purposes and in accordance with all applicable Federal Laws.

Introduction Copyright 2018 by Roger Chambers

Self Reliance Books

Get more historic titles on animal and stock breeding, gardening and old fashioned skills by visiting us at:

http://selfreliancebooks.blogspot.com/

Introduction

I am pleased to present yet another title in our "How To ..." series.

The work is in the Public Domain and is re-printed here in accordance with Federal Laws.

As with all reprinted books of this age that are intended to perfectly reproduce the original edition, considerable pains and effort had to be undertaken to correct fading and sometimes outright damage to existing proofs of this title. At times, this task is quite monumental, requiring an almost total "rebuilding" of some pages from digital proofs of multiple copies. Despite this, imperfections still sometimes exist in the final proof and may detract from the visual appearance of the text.

I hope you enjoy reading this book as much as I enjoyed making it available to readers again.

Roger Chambers

PREFACE.

THIS little book is intended strictly as an *introduction* to the science of which it treats. The Education Department requires that the teaching of Elementary Science in the schools under its direction should be " purely descriptive and experimental," and I have here attempted to indicate how such teaching should be carried on. Our *theories* may and do frequently change, but the *facts* of Nature upon which they are founded are immutable. Let us study the facts, and the theories will adjust themselves.

Each and every chapter of this book has been given as an object-lesson many times to classes of children averaging sixty in number, and of ages from ten to sixteen. Every encouragement should be given to young students to experiment on their own account; and it will be found that they are wonderfully eager to do so, and also that many of them are very apt in the construction of simple apparatus. Care has been taken to describe mostly such experiments as any student may repeat at home, with but little expenditure of money.

W. J. H.

BIRMINGHAM, *January 1895.*

CONTENTS.

    I. ELECTRICAL ATTRACTIONS, ... 7
   II. ELECTRICAL ATTRACTIONS—*Continued*, ... 13
  III. DEVELOPMENT OF ELECTRICITY, ... 18
  IV. ELECTRICAL REPULSIONS, ... 24
   V. THEORY AND LAWS OF ELECTRICITY, ... 27
  VI. ELECTRICAL STATES—GOLD-LEAF ELECTROSCOPE, ... 31
 VII. ELECTRICAL INDUCTION, ... 38
VIII. THE ELECTROPHORUS, ... 45
  IX. DISTRIBUTION OF ELECTRICITY, ... 49
   X. ELECTRICAL MACHINES, ... 56
  XI. ELECTRICAL MACHINES—*Continued*, ... 62
 XII. ELECTRICAL CONDENSERS, ... 67
XIII. THE LEYDEN JAR, ... 72
XIV. EFFECTS OF ELECTRICITY, ... 79
 XV. ATMOSPHERIC ELECTRICITY, ... 85

APPENDIX.

EXAMINATION QUESTIONS SET BY H.M. INSPECTORS OF SCHOOLS, ... 94
APPARATUS REQUIRED FOR EXPERIMENTS IN FRICTIONAL ELECTRICITY, 95

FRICTIONAL ELECTRICITY.

I.—ELECTRICAL ATTRACTIONS.

1. Origin of the Word "Electricity"—2. History of Electricity—3. Electrical Substances—4. How to Experiment—5. Attraction by Rubbed Sealing-Wax—6. Attraction by Rubbed Glass—7. Attraction by other Substances.

1. Origin of the Word "Electricity."—About the year 600 B.C., Thales, a Greek philosopher, recorded the fact that when amber was rubbed with wool it was able to attract, or draw towards it, any light bodies, such as feathers, bits of straw, etc. Now the Greek name for amber was *elektron*, and to the force which enables amber and other bodies to act in this way the name of *electricity* was applied, because it was first noticed as a property of amber. This amber is a yellow substance used for making pipe-stems, beads, etc. It is a kind of gum or resin, which has become hardened by lying in the crust of the Earth for many thousands of years. Large quantities of it are now obtained from certain sandy beds which form the southern shore of the Baltic Sea.

2. History of Electricity.—During a period of

more than two thousand years—from 600 B.C. to 1600 A.D.—hardly any discoveries were made about electricity; but in the latter year a famous physician, Dr. Gilbert, published a book which contained many new facts. Then Stephen Gray (died 1736) discovered that electricity could *pass along* certain substances; Benjamin Franklin (died 1790) proved that lightning was an electrical phenomenon; while the Italian professors Galvani and Volta showed, about the year 1800, that electricity could be produced by chemical action. In later years, Michael Faraday (died 1867) added much to our knowledge of the facts of electricity.

3. Electrical Substances.—We have seen that amber was the first substance which was found to possess electrical properties. Gilbert showed that glass, gems, sulphur, resin, and many other substances, possess the same power as amber. In our experiments we shall also use *ebonite* or *vulcanite*, a hard, black substance made by strongly heating a mixture of india-rubber and sulphur; and *sealing-wax*, which is composed of shellac and Venice turpentine, being coloured red by a little vermilion, or black by lamp-black.

Other bodies, such as the metals, wood, etc., were once thought to be incapable of electrification; but this was a mistake, as we shall presently show. Still, for ordinary experiments, it is more convenient to use substances such as glass, sealing-wax, etc., which can be electrified by friction while they are held in the hand.

4. How to Experiment.—Before commencing any

experiment, see that you have at hand all the articles which you will require. Read over carefully and repeatedly the description of the experiment which you are about to perform. For experiments in *frictional* electricity (in which we obtain electricity by the friction, or rubbing, of two substances one against the other), both the rubber and the substance rubbed must be thoroughly *dry*. To obtain this dry state, it is usual to heat or warm the bodies employed, since heat drives away moisture. On a dry day and with a bright fire at hand, every experiment ought to succeed; but in damp weather, and in rooms without a fire, even the most skilful experimenter may fail. A hot plate is very useful. It is simply a plate of iron about two feet long and one foot wide, having the edges turned up, being supported by a tripod stand. A Bunsen gas-burner, or a spirit-lamp, will soon heat the iron plate; and the rubbers, and indeed all the articles to be used in the experiments, should be kept on the top of the hot plate, and made as warm as possible. Still, we must remember that heat only acts by evaporating the water and thus procuring *dryness*.

Every experiment must be *repeated* several times, until it can be performed with neatness and with certainty. It is not sufficient to *see* an experiment performed; every student must handle the apparatus and use it for himself. Much of the apparatus is simple, and can be made by each student. When a piece of apparatus has been successfully made and used, its principles are certain to be understood and remembered.

5. Attraction by Rubbed Sealing-Wax.—Arrange upon a table little heaps of light bodies, such as bran, small feathers (down), bits of tissue-paper, fragments of gold-leaf, and a few pith-balls. The pith-balls are made out of the pith of the elder tree, which is dried, and then rounded by the help of a sharp knife and some emery-paper into little balls about a quarter of an inch in diameter. Balls of cork may be used if pith-balls cannot be obtained.

Now smartly rub a stick of dry sealing-wax with a hot, dry flannel, and hold the rubbed part near the light bodies. They will each and all fly to it as if drawn by invisible threads (Fig. 1). Indeed, this was one reason which the ancients assigned for the attractive power of amber. They said it threw out fine sticky threads, which stuck to the light bodies and drew them towards the amber. But if we place a thin sheet of glass *between* the rubbed wax and the light bodies, we shall find that they are still attracted, and it is certain that no such threads could pass through the glass. No; there must be some power, some force in or upon the sealing-wax by which it is able to draw these substances towards itself; and to this force the name of Electricity has been given.

Fig. 1.—Attraction of light bodies by sealing-wax.

6. Attraction by Rubbed Glass.—Almost any piece of glass will answer for this experiment if it be made dry by heat, but the best form to use is a glass tube 18 inches long by 1 inch broad, closed and rounded at one end and corked at the other; the inside of the tube should be kept perfectly clean and dry. The best rubber is a piece of good black silk one foot square, doubled twice, so as to have four thicknesses, and sewn across and across with silk thread. The electricity is produced more easily if some electric amalgam is smeared on the silk rubber. This amalgam is made by melting in an iron spoon equal weights (say half-ounce of each) of tin and zinc, and then gradually adding twice as much by weight (two ounces) of warm, dry mercury. This amalgam is a brittle mass, which must be pounded in a mortar, and made to adhere to the silk by the help of a little pure lard. When the air is very dry, as on a frosty day in spring with an east wind, very light bodies may be attracted at a distance of a foot or more by the electricity produced by briskly *exciting* (or rubbing) a glass tube. The tube must be held by the other extremity while the rounded end is briskly and smartly rubbed with the silk, which should surround the tube. Notice that it is only the rubbed part of the glass which has the power of attraction.

7. Attraction by other Substances.—A vulcanite penholder or ruler, rubbed with warm flannel, acts almost better than a stick of sealing-wax, since it is not so brittle. Shellac is easy to electrify in the same way, and the expense of a solid stick may be saved by making a tight roll of brown paper and then covering it thickly with shellac dissolved in spirits of wine (methylated spirit will do).

Other examples of bodies readily electrified by friction are:—(1.) A piece of resin rubbed on hot dry flannel. (2.) Dry hot sheet of brown paper brushed smartly with clothes-brush: the brown paper will not only attract light bodies, but will cling to any wall against which it is placed. Instead of being brushed it may be drawn smartly between the knees three or four times. (3.) Thin (foreign-post) writing-paper laid on a hot, dry mahogany board and stroked with india-rubber. (4.) Collodion is a liquid substance used by photographers; it is made from gun-cotton. When collodion is spread out into a thin layer or film, as by pouring it upon a sheet of glass, it soon dries. Collodion balloons can be made by pouring collodion into a round glass bottle (a Florence flask answers well), shaking it so as to coat the sides of the bottle, leaving it to dry, and then separating it from the glass. Simply stroking one of these collodion balloons with the dry hand is sufficient to strongly electrify it.

Any of these substances (and many others which we have not named) exhibit electrical properties when rubbed, the most noticeable being that power of attracting light substances which so excited the wonder of the ancients.

II.—ELECTRICAL ATTRACTIONS.
(Continued.)

8. The Electric Pendulum—9. The Balanced-Straw Electroscope—10. The Balanced Lath—11. Attraction of Excited by Unexcited Bodies—12. Construction of Balanced Tubes and their Supports—13. Attraction of Electrified Bodies by each other.

8. The Electric Pendulum.—Hitherto the light bodies which were attracted by electrified rods, etc., have been simply allowed to lie on any flat surface. For ordinary experiments the pith-ball electroscope, or electric pendulum, is more convenient. In Fig. 2 we see a bent piece of glass tubing, or of brass, fastened to a wooden base; from this a pith-ball is suspended by a fibre of floss silk. The word "electroscope" is derived from *elektron*, electricity, and *skopeo*, I see or detect; and the instrument is so named because, by its aid, we can tell whether any substance which

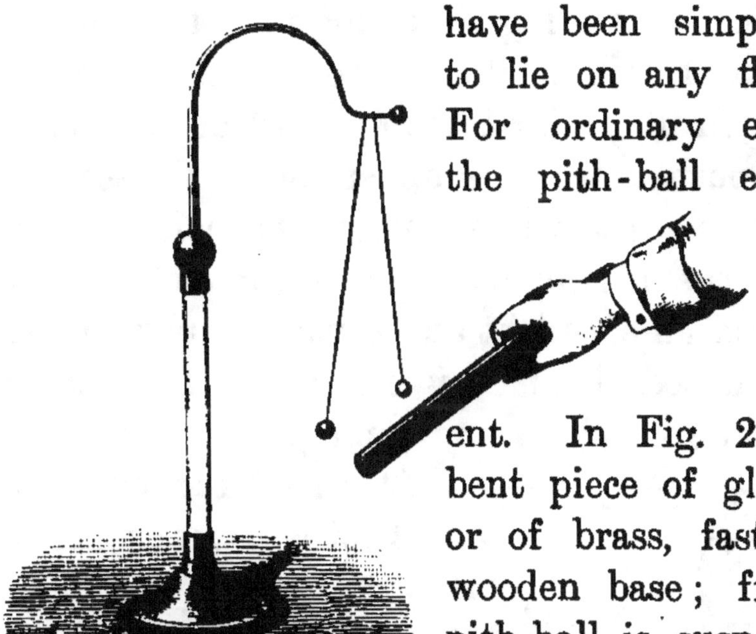

Fig. 2.—The Electric Pendulum.

we bring near the pith-ball is or is not electrified. For example, a glass tumbler in its ordinary state has no power over the suspended pith-ball; but when the same tumbler is rubbed with a clean and dry silk handkerchief, it is able to attract the pith-ball from a distance of several inches.

9. The Balanced-Straw Electroscope.—Take a strong, clean piece of straw about twelve inches in length, cut off a piece about half-an-inch long from one end, and fasten this to the middle of the straw by the aid of a little sealing-wax. Slit each end of the straw and insert pointers of coloured paper. Now make a stand by cutting a piece of tin or zinc about three inches square; to the centre of the tin fasten an upright piece of sealing-wax three or four inches long (scratch the tin in the centre to make the hot wax adhere firmly); heat the eye end of a sewing-needle, and push it into the upper end of the sealing-wax. Then place the straw upon the point of the needle, and the instrument will be complete. This will be found an excellent and sensitive instrument; an electrified body held near either end of the straw ought to make it spin round rapidly.

FIG. 3.—Balanced Lath.

10. The Balanced Lath.—Another means of exhibiting electrical attraction is to balance a smooth lath, three or four feet in length, upon some smooth, round surface, as a watch-glass, an egg in an egg-cup, or the top of an inverted Florence flask supported by a bottle (Fig. 3). When an electrified body is brought near to either end of the lath, the latter will move towards it.

11. Attraction of Excited by Unexcited Bodies.— An electrified body is able to influence and to attract any substance to which it may be brought near. If the substance is very light, or if it is balanced so as to move easily, then motion will result; but in any case, whether there be visible motion or not, all known substances are influenced by the presence of an electrified body. But is there any reaction? Do the substances which are affected react upon the excited body, and attract it in turn? To prove this, it is only necessary to excite a piece of sealing-wax and place it in a little wire stirrup, as shown in Fig. 4, the whole being suspended by a narrow silk ribbon. Let some substance which has *not* been rubbed — as the finger, a pencil, a book, etc.—be brought near the rubbed part of the wax; instantly the excited wax will move towards the unexcited body. This proves that electrified substances are attracted, or influenced, by bodies which have not been electrified. It is, in fact, a proof of Newton's third law of motion, which states that "action and reaction are equal and opposite:" electrified bodies attract neutral substances, and neutral substances attract electrified bodies. The experiment may be repeated with an excited glass rod, when exactly the same result will

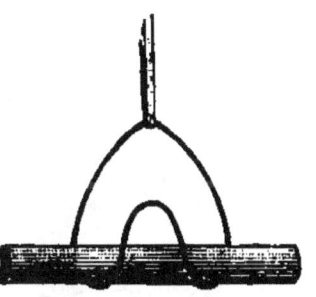

Fig. 4.—Excited wax suspended in wire stirrup.

be obtained—the excited glass will move towards any neutral body which may be brought near it.

12. Construction of Balanced Tubes and their Supports.—To balance a glass tube, one end should be closed and rounded, and the other end all but closed. Now find the centre (by balancing the tube on a knife-edge), and mark it by a scratch with a file, or by a spot of ink. Heat the centre strongly, and push in the soft glass with the pointed end of a lead pencil. A support for the tube can be made by cutting off the end of a shawl-pin and forcing the blunt end (with the help of a pair of pliers) into a bung; or a darning-needle similarly pushed into the cork of any large stout bottle answers equally well (see Fig. 5). To balance a stick of sealing-wax, a glass pivot about a quarter of an inch long must be made by drawing out and rounding a piece of glass-tubing in a gas-jet. The pivot can then be pushed into the centre of the stick of wax by warming the glass and placing a hot wire inside it.

Fig. 5.—Glass tube balanced on pivot.

13. Attraction of Electrified Bodies by each other.—Excite a glass tube with amalgamed silk, and suspend or balance it as before. We know that it will be attracted by a stick of unexcited wax. But now excite the wax by briskly rubbing it with dry flannel. On bringing the excited wax near the excited glass, the attraction will be found to be *much greater* than before. In the same way excited glass *strongly* attracts excited wax. It is the same with many other substances. We find that—

Glass rubbed with silk attracts and is attracted by wax rubbed with flannel.

India-rubber rubbed with paper attracts and is attracted by sulphur rubbed with flannel.

Amber rubbed with flannel attracts and is attracted by glass rubbed with silk.

Vulcanite rubbed with flannel attracts and is attracted by glass rubbed with silk; etc., etc.

III.—DEVELOPMENT OF ELECTRICITY.

14. Electrics and Non-Electrics—15. Conductors and Non-Conductors—16. Table of Conductors and Non-Conductors—17. All Bodies can be Electrified by Friction—18. The Human Body Electrified by Friction—19. Why Moisture interferes with Experiments in Frictional Electricity.

14. Electrics and Non-Electrics.—Until about a century and a half ago, it was supposed that only certain substances, such as amber, glass, sealing-wax, etc., were capable of electrification. The reason for this idea lay in the fact that it is impossible, by any amount of friction, to get signs of electricity from most substances, such as metals, wood, etc., if they are rubbed while *held in the bare hand*. Thus, suppose we take a brass tube, and holding part of the brass in one hand, rub or flap another part of the brass with a piece of flannel or fur, we shall never get the brass to attract light bodies, as a glass tube would do if similarly held and rubbed; no, not if we rubbed until the brass wore away. For this reason substances such as amber, glass, wax, etc., were called *electrics;* while all the metals, wood, etc., were called *non-electrics*, because it was supposed that they could not be electrified.

15. Conductors and Non-Conductors. — In 1729 Stephen Gray discovered that electricity could pass along a thread of cotton, but not along a thread of silk. He found that when a cork, or a long piece of wood, was inserted in a glass tube, that upon rubbing the glass the distant end of the cork or of the wood was electrified; or, if a bullet was hung by a cotton thread to an excited glass tube, the electricity passed downwards to the bullet, which was then able to attract any light substance (Fig. 6).

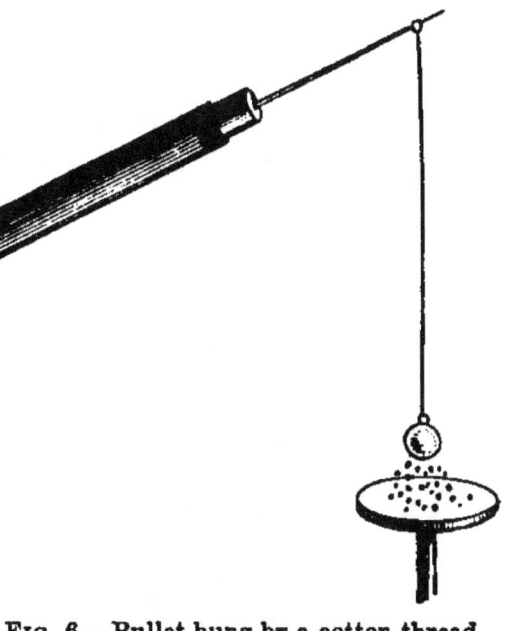

FIG. 6.—Bullet hung by a cotton thread from excited tube attracts light bodies.

A classification of substances was soon made: all bodies which refused to allow electricity to pass along them were called *non-conductors*, while to others which appeared to offer no resistance to the passage of electricity the name of *conductors* was applied. In the following table the *best* conductor is placed first, and the best non-conductor (or *worst* conductor) is put last.

It has since been found that there is no *perfect* conductor; that even silver and copper, which are the best conductors, offer a *little* resistance to the passage of electricity. Neither is there any *perfect* non-conductor; for glass, wax, shellac, etc., allow a little of the electricity to pass, though but very slowly.

16. Table of Conductors and Non-Conductors:—

Silver, Copper, Other Metals, Salt Water,	Good Conductors.
Pure Water, Bodies of Animals, Cotton, Linen, Hemp, Dry Wood, Paper, Ice,	Partial Conductors.
Fats and Oils, Porcelain, Wool, Silk, Gutta-percha, Ebonite, Wax, Shellac, Sulphur, Amber, Glass, Paraffin, Dry Air,	Non-Conductors, or Insulators.

Non-conductors are also called insulators (Latin *insula*, an island), or isolators. Faraday called them *dielectrics*, because they serve to keep the two kinds of electricity apart. Flames and hot-air currents are good conductors of electricity. If a rubbed rod of glass or vulcanite be passed through the flame of a spirit lamp, it is immediately and completely *discharged*—that is, it loses all its electricity.

17. All Bodies can be Electrified by Friction.—If we fully understand Gray's discovery, we know why it is that most bodies cannot be got to show any signs of electricity if they are held in the hand while they are being rubbed. For, if any electricity *is* produced on the brass tube already referred to by rubbing it with flannel, it is certain that this electricity will spread all over the brass, and thence through the body of the person holding the brass, into the earth, which is able to hold any amount of electricity. The brass is a very good and the human body a fairly good *conductor*. But if we *interpose* a non-conductor between the conductor and the earth, then we shall insulate, or isolate, the conductor; any electricity produced upon the conductor will then be unable to escape, and we shall be able to examine it at our leisure. For this purpose it will only be necessary to fit a handle to the brass tube (Fig. 7), composed of  some insulating substance, such as glass or vulcanite. Holding the tube by this handle, and taking care to keep our fingers as far away from the brass as possible, let us smartly *flap* the brass with a piece of dry flannel, or, better still, with a fox's tail or a prepared cat-skin. On now bringing any part of the brass tube near the pith-ball electroscope, an instant attraction of the pith-ball will be obtained.

FIG. 7.—Brass tube with insulating handle.

This experiment may be repeated with all the other substances which were once called "non-electrics," and it will be found that when they are properly insulated, all these substances can be electrified by friction. Thus, we may take an apple between our fingers and flap it as long as we like without its showing any signs of excitation. But tie a dry silk thread to the stalk of the apple, and, holding the end of the thread, proceed to flap the apple with a fox's brush as before; after a few strokes the apple will be able to cause light bodies to fly towards it, to attract pith-balls, and, in fact, to show all the signs of electrification. For this reason the terms "electrics" and "non-electrics" are now no longer used, since all bodies are now known to be capable of electrical excitation. Those bodies which were formerly called "electrics" are now styled *non-conductors*, while the "non-electrics" are known as *conductors*.

18. The Human Body Electrified by Friction.—When a non-conductor is rubbed, the electricity *remains on the part rubbed*; but when a conductor is rubbed, the electricity produced *spreads all over* the surface of the body, and escapes thence, if it be possible, to the earth. It is easy to electrify the human body by friction. Let a boy stand in the middle of a room upon a board supported by four dry glass tumblers, and let a second boy flap his back smartly with a fox's tail. In a few moments sparks may be drawn from any part of the body of the insulated boy. The experiment succeeds better if the boy standing upon the board or insulating stool wears a mackintosh. Instead of the friction with fur, the same effect will be produced if a vulcanite comb is repeatedly passed by some second person through the boy's hair.

Not only solids but liquids and gases also can be electrified by friction.

19. Why Moisture interferes with Experiments in Frictional Electricity.—Since water is a conductor of electricity, it is evident that it must be difficult, or impossible, to cause damp substances to show any signs of electricity. If a wet or even a moist glass tube be rubbed ever so energetically with a piece of damp silk, no attractive power will be shown by the glass. The reason is plain: electricity was produced on the glass, but as fast as it was produced it escaped along the layer of water which coated the tube to the body of the person holding it, and thence to the earth. Now let the glass rod and the silk rubber be each thoroughly heated until all the water has been made to evaporate: then every stroke of the silk upon the glass will excite the latter, a crackling noise will be heard, the glass tube will strongly attract light bodies, and if the experiment is performed in the dark, flashes of light will be seen. Glass is a substance which is naturally more or less damp, because it condenses upon its surface the water-vapour contained in the air, as we may see on our window-panes: the term *hygroscopic* (detecter or attractor of water) is applied to such substances. Shellac, vulcanite, etc., do not act in this way, and it is consequently a good plan to coat all the glass legs or supports used for electrical apparatus with shellac varnish.

IV.—ELECTRICAL REPULSIONS.

20. How to obtain Repulsion—21. Two Kinds of Electricity—22. Vitreous and Resinous Electricity—23. Positive and Negative Electricity.

20. How to obtain Repulsion.—So far, all the electrical phenomena we have witnessed have been examples of attraction. We have seen (1) that an electrified body can attract *any* unelectrified body; (2) that an electrified body can attract certain other electrified bodies, as when polished glass excited with silk is brought near to sealing-wax excited with flannel. We have now to learn that the force of electricity can produce *repulsion* as well as attraction. For this purpose take two glass tubes and rub them simultaneously with amalgamed silk. Now suspend or balance one tube, and bring the second tube near it: strong repulsion will at once be observed, and the balanced tube will *move away* from the other, although a distance of several inches may separate them. Repeat the experiment, using two sticks of sealing-wax excited with flannel instead of the two glass tubes. It will be found that excited wax repels excited wax. Generally it will be found that any two *like* substances which have been excited by friction with the same substance will repel each other.

21. Two Kinds of Electricity.—But we know that electrified bodies do not always repel one another. Let a pith-ball, suspended by a silk thread (Fig. 8), be touched by an excited glass rod; if, now, the same or any other similarly excited glass rod be brought near the pith-ball, the latter will be *repelled*. But let excited wax be tried, and it will be found to strongly *attract* the electrified pith-ball. From this it seems clear that there must be *two kinds* of electricity. This discovery was first made about the year 1735, by two experimenters named Symmer and Du Fay. They tried many substances in turn, and found that when electrified they all behaved either (1) like excited glass, or (2) like excited sealing-wax.

FIG. 8.—Electrified suspended pith-ball repelled by excited glass.

22. Vitreous and Resinous Electricity.—The word *vitreous* means glassy, and to the electricity produced on glass after friction with silk Du Fay gave the name of vitreous electricity, while that produced on sealing-wax by friction with flannel he named resinous electricity. But these terms are no longer used, and for a very good reason. It has since been found that the kind of electricity produced on any substance by friction is not always the same, but depends partly on the nature of the substance, partly on the state of its surface, and partly on the substance with which it is rubbed. If we rub polished glass with *fur*, we shall find what was called "resinous" electricity produced upon a vitreous surface, the proof being that glass excited in this way will *repel* sealing-wax excited with flannel. Or again, *ground* glass rubbed with flannel yields resinous electricity. On the other hand, if a tube of gutta-percha is rubbed with flannel, it will, we know, develop resinous electricity on its surface; but if it be rubbed with gun-cotton (or with collodion) then "vitreous" electricity will be found upon the resinous surface. Certain resins become vitreously electrified when rubbed with cotton; and even sealing-wax yields vitreous electricity when rubbed with chamois leather on which has been spread a soft amalgam made by rubbing together (in a mortar) tinfoil and mercury. Thus the use of the words vitreous and resinous as applied to the two kinds of electricity has been wholly given up, since it is proved by experiment that we can obtain vitreous electricity on a resinous surface, and resinous electricity on a vitreous surface.

23. Positive and Negative Electricity.—The famous American, Benjamin Franklin, argued that there was only *one* kind of electricity, and that this was of the nature of a fluid, because it could *flow* from place to place just as liquids and gases (fluids) do. He believed that a body only showed signs of electricity when it had either *more* of its usual quantity of this fluid (being then said to be *positively* electrified), or when it had *less* than its normal quantity (when it was said to be *negatively* electrified). This *one-fluid* theory of Franklin's has not been adopted generally by electricians, but his names of "positive" and "negative" have come into general use to indicate the two electrical states which certainly do exist.

V.—THEORY AND LAWS OF ELECTRICITY.

24. Two-Fluid Theory of Symmer and Du Fay—25. Properties of the Electrical Fluids—26. Use of a "Theory"—27. Law of Inverse Squares—28. Law of Quantity.

24. Two-Fluid Theory of Symmer and Du Fay.—Some time back a famous electrician, Professor Tyndall, set the following question at an examination, "Can you feel, smell, see, taste, or weigh what is called *electricity?* If not, how do you know that it exists?" The answer to this must be that we cannot detect electricity itself by means of our senses; and it has no weight, because it is not *matter*, but a force, or form of energy. We know that there *is* such a force as electricity, because we see it produce certain *effects* upon matter; for example, we see this force called electricity set bodies in motion. But again, if we are asked what *is* this force of electricity, and in what way does it produce these effects, we can only say *we do not know*. Yet it is very convenient to think of electricity as a fluid, to employ this idea as *a theory* on which to explain and classify the facts we observe.

25. Properties of the Electrical Fluids.—The two-fluid theory of electricity states:—

(1.) That every body contains an unlimited quantity of each of two electrical fluids, to which the names *positive* and *negative* may be applied. The type of "positive" electricity is that produced upon glass rubbed with silk; while by "negative" electricity we mean the same kind as that produced upon sealing-wax by friction with flannel.

(2.) That these two fluids are opposite to each other in all their properties.

(3.) That these fluids are self-repulsive, but mutually attractive—that is to say, positive repels positive, and negative repels negative; but positive attracts negative, and negative attracts positive. Putting this into as few words as possible, we may say that,—

Like electricities repel;
Unlike electricities attract.

(4.) An "unelectrified" body—that is, one which shows no signs of electricity—contains an *equal* quantity of each of the two electric fluids.

Instead of the word "positive" we frequently employ the sign + (plus), and for "negative" we write − (minus).

26. Use of a "Theory."—This two-fluid theory of electricity has held its ground, more or less, since the time when it was first stated (about the year 1735) by Symmer and Du Fay; and although the best electricians of to-day tell us that electricity is certainly *not a fluid*, yet as they cannot give us any simpler theory to put in its place, it may still be well to *conceive* of electricity as a fluid, so that we may have something to think of, some idea round which to group our facts.

THEORY AND LAWS OF ELECTRICITY. 29

27. Law of Inverse Squares.—In experiments upon the action of electrified bodies on one another, it is necessary to bring the substances *near to one another*, in order to obtain any effects. This proves at once that the force exerted diminishes as the distance increases. The result of a great number of experiments, conducted by skilled electricians with the most delicate instruments, proves that *the force of attraction or of repulsion exerted by any electrified body varies inversely as the square of the distance.* This statement is known as the law of inverse squares. We may illustrate it by supposing an excited glass tube to attract a pith-ball with a certain force at a distance of eight inches; then if we halve the distance (making it four inches) the attraction will be *four times as great;* so also the attraction at a distance of two inches will be sixteen times, and at one inch sixty-four times as great as at a distance of eight inches.

Let us work out the last statement, premising that by the *square* of a number we mean the number multiplied by itself, while we *invert* a number by making the numerator and denominator change places, as 5 inverted $= \frac{1}{5}$.

Let Attractive Force at 8 inches $= 8 \times 8 = 64$, inverted $= \frac{1}{64}$
Then Attractive Force at 1 inch $= 1 \times 1 = 1$, inverted $= 1$

It is the same when a body is *repelled* by another body similarly electrified: the repulsion at a distance of three inches is only *one-ninth* as great as at a distance of one inch; and it is only *one-hundredth* as much at a distance of ten inches as at a distance of one inch. This law of inverse squares may be proved with an instrument invented by Coulomb, and known as the torsion balance.

28. Law of Quantity.—But the force with which a substance is attracted or repelled by any electrified body depends on something more than simply the distance between them. The body may be slightly or it may be strongly electrified, and the effect it produces will vary accordingly. *The force of attraction or repulsion exerted by any electrified body varies directly with the quantity of electricity with which it is charged.* Thus, if we double the charge imparted to a body, we also *double* its power of attracting or repelling; if the charge of electricity is made four times greater, then the force will also be four times as great if the distance remain the same.

VI.—ELECTRICAL STATES—GOLD-LEAF ELECTROSCOPE.

29. Separation of the Electric Fluids—30. Insulation of the Rubber—31. Classification of Positive and Negative Substances—32. Friction develops Equal Quantities of the Two Electric Fluids—33. Construction of the Gold-leaf Electroscope—34. Use of the Gold-leaf Electroscope—35. The Gold-leaf Electroscope as a Test for Conductors—36. The Proof-Plane.

29. Separation of the Electric Fluids.—When a body shows no signs of electricity, it is said to be in a *neutral* state. We believe that it then has an equal amount of positive and of negative electricity, and that these two fluids neutralize one another. If we can take away some of the negative electricity, then there will be left an excess of positive electricity; on the other hand, if some of the positive electricity be removed, the body will show signs of, or, as we say, will be *charged* with, negative electricity.

Consider the state of a glass tube and of a silk rubber, in their neutral state, before they touch one another. Neither glass nor silk shows any signs of electricity, because they each contain an equal quantity of the two electric fluids. But now rub the two bodies together; then by the friction some positive electricity is caused to pass from the silk to the surface of the glass, while negative electricity passes from the glass to the silk, and each of the two substances can then be made to give signs of electrification. Thus when any two unlike substances are rubbed together, they will each be electrified, the one positively and the other negatively. It may be objected that the silk rubber does not generally show any signs of the negative electricity which we have stated passes to it when it is rubbed against glass. But we must remember that the silk rubber is usually held in the hand, and the negative electricity consequently passes to the earth through the body of the person holding the rubber as fast as it is produced upon or passes from the glass to the silk. If the rubber is *insulated*, so that its electricity cannot escape, then in every case it will be found—

(1.) That both the substance rubbed and the rubber are electrified.

(2.) That the electricity upon the rubber is always *opposite* in kind but *equal* in amount to that upon the body rubbed.

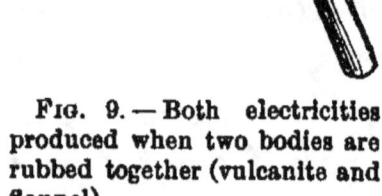

Fig. 9.—Both electricities produced when two bodies are rubbed together (vulcanite and flannel).

30. **Insulation of the Rubber.**—To prove that electricity is produced upon the rubber, we may cover a long cork with silk, pass an ebonite penholder (to serve as an insulating handle) through the cork, and smartly rub the silk upon a hot, dry strip of glass. Both glass and silk will be found to be electrified—the glass positively, and the silk

negatively. Or a strip of flannel may be wrapped round a rod of vulcanite, and then rubbed up and down the vulcanite by means of silk threads tied round the flannel (Fig. 9); when the two bodies are separated, the vulcanite will be found to be negatively and the flannel positively electrified.

31. Classification of Positive and Negative Substances.—In the following list the substances are arranged so that the most electro-positive is placed first. If any two of the substances named below be rubbed together, the one which stands highest on the list will be electrified positively, and the other negatively.

+
1. Fur of Animals.
2. Wool (flannel, etc.).
3. Polished Glass.
4. Silk.
5. The Hand.
6. Metals.
7. Vulcanite.
8. Sealing-wax.
9. Sulphur.
10. Gutta-percha.
11. Gun-cotton (collodion, etc.).
−

32. Friction develops Equal Quantities of the Two Electric Fluids.—To prove that the quantity of electricity developed upon the rubber is *equal* in amount to that produced upon the body rubbed, the two bodies (the flannel and the vulcanite, for instance) may be laid *together* upon a gold-leaf electroscope. No divergence of the leaves takes place until *one* of the bodies is removed. As they lie side by side upon the plate of the electroscope, the + on the flannel *exactly* neutralizes the effect of the − upon the vulcanite, to which, therefore, it must be precisely equal in quantity.

33. Construction of the Gold-leaf Electroscope.—The attraction of light bodies, or of a suspended pith-ball, is not a test sufficiently delicate to detect *small* quantities of electricity. For this purpose the gold-leaf electroscope is a more useful instrument. To make one, we require a wide glass flask, an india-rubber cork, six inches of stout brass wire, some strips of gold leaf or of Dutch metal (the latter is much easier to handle than the gold leaf), and a smooth circular piece of brass (or any metal) an inch or so in diameter. Heat the wire, and pass it through the centre of the cork; solder the brass disc to one end of the wire; bend the other end of the wire, hammer it flat, and affix a strip of gold leaf

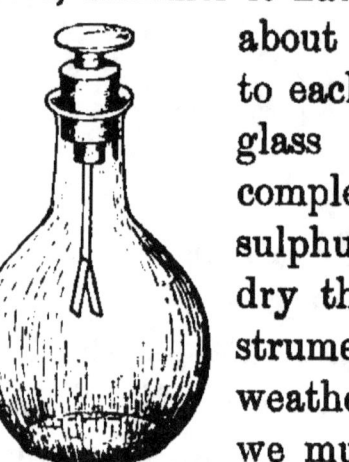

Fig. 10.—Gold-leaf Electroscope (glass flask, etc.).

about $2\frac{1}{2}$ inches long and $\frac{1}{2}$-inch wide to each side. Place the cork in the glass flask, and the instrument is complete (Fig. 10). A little strong sulphuric acid kept in the flask will dry the air inside and cause the instrument to act better in damp weather. If a common cork is used, we must bore a hole in it and insert a piece of glass tubing filled with shellac softened by spirits of wine; the wire must then be pushed through the centre of the shellac.

In the gold-leaf electroscope we have an insulated conductor extending from the brass disc to the gold leaves. We must remember that the gold leaves are extremely light and very close together, and that electricity can pass down to them with the greatest ease along the brass wire from the brass disc on top.

34. Use of the Gold-leaf Electroscope.—When in a neutral state, the two gold leaves hang vertically side by side. Now let the brass disc be *touched* with a glass tube which has been excited by friction with silk. Positive electricity will pass from the glass to the brass, and will spread all over disc, wire, and leaves. The two leaves will then repel one another, and will start apart (Fig. 11). The electroscope is

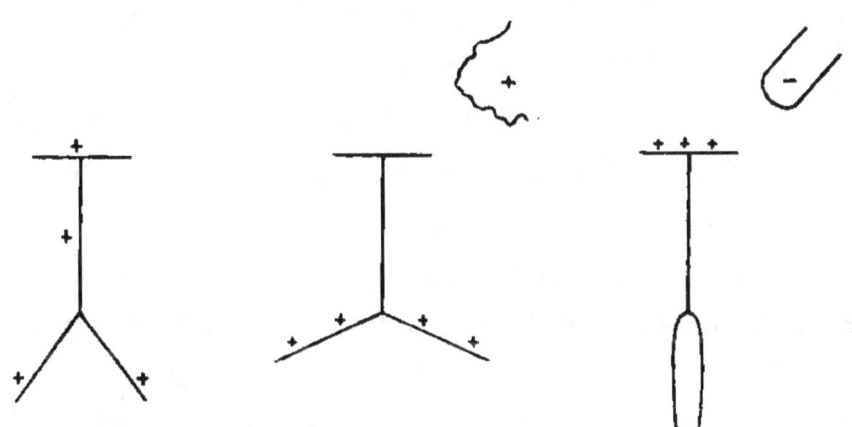

Fig. 11.—Electroscope charged with positive electricity by contact with glass rod.

Fig. 12.—State of electroscope when excited brown paper is brought near.

Fig. 13.—State of electroscope when negatively excited body is brought near.

now charged with positive electricity. If, on the other hand, we wish to charge it with negative electricity, we must touch the disc with excited wax instead of excited glass.

Let us suppose that the leaves are diverging (as shown in Fig. 11), because they each contain positive electricity. Now we wish to find out which kind of electricity is produced on brown paper that has been smartly brushed. On bringing the brown paper near the disc of the electroscope, we observe that the leaves move *further apart*. The electricity on the paper must have *repelled* electricity from the disc into the leaves (Fig. 12). Therefore the electricity on the paper must be of the *same kind* as that on the electroscope (for *like* electricities repel one another)—that is, of the *positive* kind.

If the leaves *fall together* when a charged body is brought near to the disc (Fig. 13), it is *probable* that the electricity on the body is of the opposite kind to that on the leaves, but it is not absolutely certain. For the approach of a *neutral* body will cause the leaves to act in this way—that is, to collapse; so that repulsion is the only sure test.

35. The Gold-leaf Electroscope as a Test for Conductors.—Connect an electrified body, as excited glass or vulcanite, with a gold-leaf electroscope by means of a piece of sewing-cotton: twist the cotton two or three times round a glass rod, for example, and then rub it up and down the rod by means of a piece of silk. The gold leaves quickly diverge, because cotton is a conductor, and allows the electricity to flow along it; the same effect is produced by wires, by string, and in fact by all conductors. But replace the cotton by a thread of dry silk, and no divergence of the leaves can now be obtained, because the electricity cannot pass along the silk; it is an insulator. Now wet the silk, and the leaves immediately diverge, thus proving water to be a conductor. By this means we can distinguish substances which insulate from those which conduct electricity.

An even simpler method is to charge the electroscope so that the leaves remain standing apart, and then to touch it in succession with the substances whose conducting powers it is desired to test. No change occurs when the charged electroscope is touched with sulphur, a paraffin candle, etc.; therefore these bodies are non-conductors. But the instant we use a rod of any metal the leaves collapse, because the metal offers a free passage to the ground to the electricity with which the leaves were charged. Every substance tested must be perfectly dry, for the film of water upon a damp insulator acts as a conductor.

36. The Proof-Plane.—When we require to test the electricity of a body, it is often more convenient to carry a little of the electricity by means of an instrument called a proof-plane, than to bring the body itself to the electroscope. A proof-plane consists of a small, usually flat, and circular conductor, provided with an insulating handle (Fig. 14).

Fig. 14.—Proof-Plane.

One may be made (a) by cutting a cardboard disc about two inches in diameter, gumming gilt paper upon each side, and fastening it to a strip of varnished glass; (b) by making a small tin disc with a socket, into which a piece of glass rod or tubing is fitted, or a vulcanite penholder; (c) by cutting out a metal disc (with smooth edges), and affixing a stick of sealing-wax to the centre of the disc. To use the proof-plane, we simply touch with it the body whose electricity we desire to test, and then carry the charged proof-plane to an electroscope.

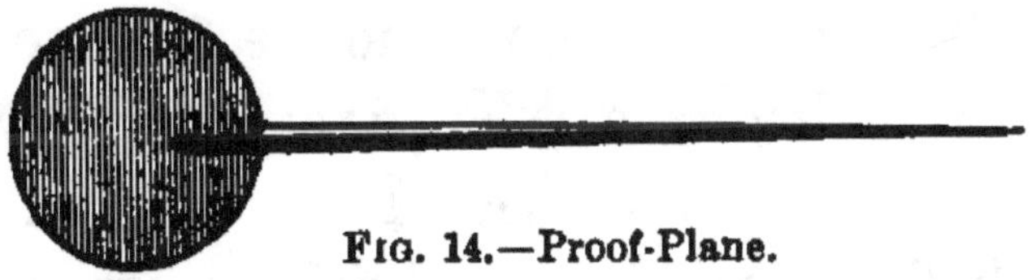

VII.—ELECTRICAL INDUCTION.

37. Action at a Distance—38. Examples of Induction—39. Entrapping Electricity by Induction—40. Gold-leaf Electroscope charged by Induction—41. Entrapping Electricity in the Gold-leaf Electroscope—42. Local Analysis—43. Cause of the Attraction of Neutral Bodies—44. Neutral Bodies are more strongly attracted when uninsulated.

37. Action at a Distance.—It is a distinguishing character of the *physical* forces, among which we reckon electricity, that they can *act at a distance.* Thus the force of gravitation acts across the millions of miles which separate the Earth from the Sun, and causes the former to revolve round the latter; the forces of light and heat come to us without any difficulty from the fixed stars, whose distances far exceed that of the Sun; and in the same way we believe that the electricity upon an excited substance can influence the electric fluids upon other bodies at a distance from it.

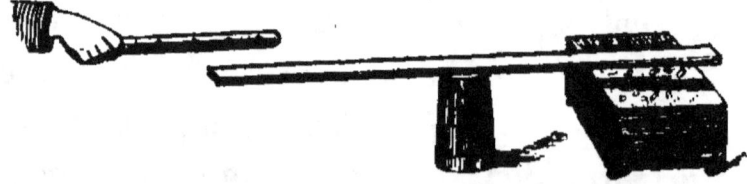

FIG. 15.—Lath on glass tumbler: hold excited vulcanite ruler over one end; other end attracts light bodies.

38. Examples of Induction.—To illustrate this inductive action of electrified bodies, we may place a lath upon an insulating support (Fig. 15), consisting of a dry glass tumbler. Let some bits of tissue-paper or of bran be placed about an inch below one end of the lath. On bringing an excited vulcanite ruler over the distant end of the lath, the bits of paper will be seen to be attracted by the end nearest to them. Keeping the vulcanite in its place, an inch or two distant from the lath, let the electricity on the other end of the lath, by which the paper was attracted, be tested by means of the proof-plane. It will be found to be *negative*. The electricity on the excited vulcanite was negative: when the vulcanite was held near the neutral lath, the electricity on the latter body was decomposed, positive being attracted to the end nearest the vulcanite, and negative repelled to the other end. When the charged vulcanite is removed the two electricities upon the lath run together again, and it then shows no signs of electricity.

To this action of electricity *at a distance* the name of *induction* is applied.

39. Entrapping Electricity by Induction.—Let the two insulated neutral brass cylinders, A B and C D (Fig. 16), be placed end to end, touching one another. Now strongly excite a glass tube, X, and hold it over the end of the cylinder, C D. What will happen? Clearly, − electricity will be drawn to the part of C D which is nearest to X, while + will

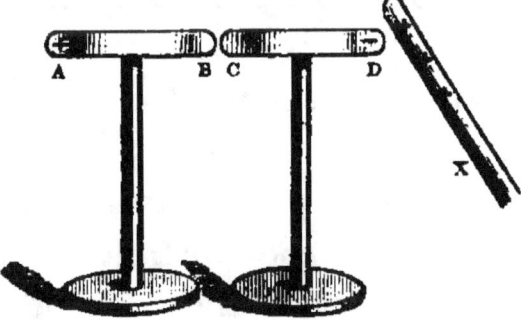

FIG. 16.—Insulated brass cylinders in contact; excited glass tube held over one end.

be repelled into A B. A portion of the + electricity on A B may now be removed by a proof-plane and tested; but if we attempt to get any − electricity by touching C D, we shall fail, for that − electricity is held fast, is *bound* as it were, by the attraction of the + upon X.

But, without touching the brass cylinders, and while they are under the influence of X, let them be separated by removing A B. Now take away the charged glass rod, and *each* of the cylinders will be found to show signs of electricity—A B will be +ly and C D −ly electrified. This electricity has not been produced by friction, but it has been entrapped, as it were, by the inductive influence of X. The electrical condition of the cylinders may be made evident by suspending pairs of pith-balls, by cotton threads, from each end of each cylinder. Where there is free electricity the pith-balls will repel each other, and will stand apart.

40. Gold-leaf Electroscope charged by Induction.— In a previous experiment we charged a gold-leaf electroscope by actually imparting to it, by contact, either positive or negative electricity. But it is possible to charge the instrument without touching it with any electrified body.

Let a +ly electrified glass tube be brought near (say two inches distant from) the disc of the electroscope; then the state of the instrument is represented in Fig. 17: the negative electricity accumulates in the disc, while positive is repelled into the leaves, which diverge widely. Now let any person touch the disc with the finger, or with any conductor connected with the earth, and the leaves collapse (Fig. 18). What has happened? Clearly the positive electricity has rushed up from the leaves and escaped through the person's body to the ground.

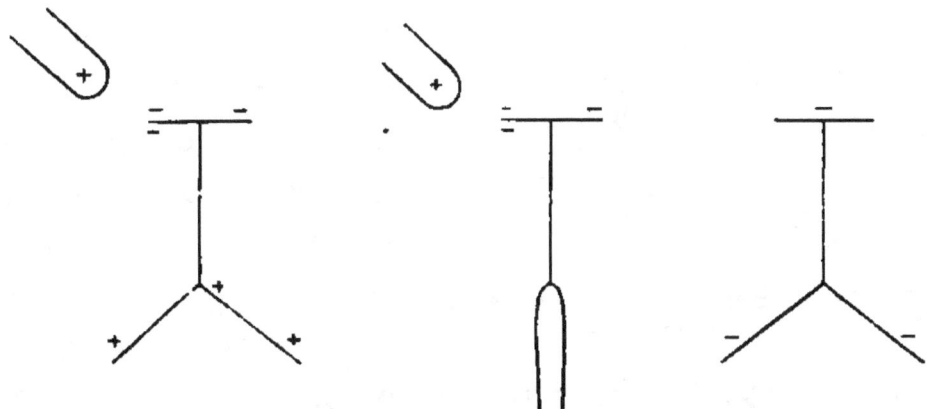

FIG. 17. — Positively electrified glass tube brought near disc of electroscope.

FIG. 18.—State of electroscope when finger is placed on disc.

FIG. 19.—State of electroscope when glass tube and finger are removed.

Now remove the finger, and then, last of all, take away the influencing body, the +ly electrified glass tube; then the gold leaves once more diverge, because the negative electricity now distributes itself equally all over disc, wire, and leaves, and the leaves consequently again repel one another (Fig. 19). When an electroscope is charged by induction in this manner, we must remember that its electricity is always of the *opposite* kind to that of the inducing body.

41. Entrapping Electricity in the Gold-leaf Electroscope.—But it is possible, by induction, to charge an electroscope with the *same* kind of electricity as that which resides upon the inducing body. Let a gold-leaf electroscope be connected, by means of a copper wire, with an insulated brass ball (Fig. 20). Now bring any electrified body, as excited vulcanite, near the ball; then + electricity will occupy the ball, while the − will be repelled into the electroscope and will cause the leaves to diverge. Now remove the connecting wire by means of some insulating substance (this may be done with the fingers if pieces of stout rubber tubing be slipped over them), and then take away the excited vulcanite. The brass ball will be left charged positively, and the electroscope negatively. It will be seen that this experiment is similar to that described in paragraph 39.

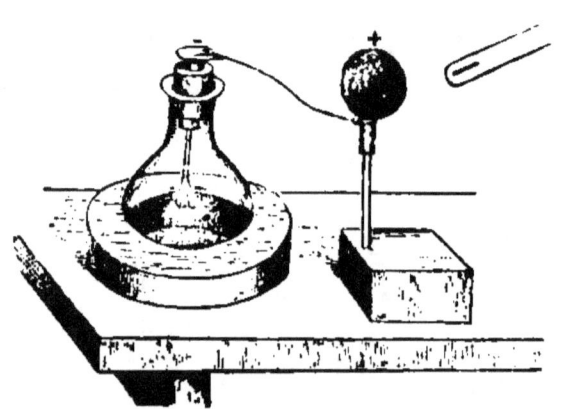

FIG. 20.—Entrapping electricity in gold-leaf electroscope.

42. Local Analysis.—We now see that an electrified body can decompose, or analyse, the electricity of any substance to which it may be brought near. *Every* substance around us may be, ordinarily, considered as in a neutral state, containing an equal quantity of each of the two electric fluids. If, therefore, we bring an electrified body near any neutral substance, it will attract the opposite fluid to that with which it is charged, and will repel the like kind. Thus if we bring a rod of excited glass near to a neutral insulated brass ball, we shall be able, so long as the excited glass is held near the ball, to take + electricity by means of a proof-plane from the side of the ball farthest from the excited glass.

Hold a stick of excited sealing-wax about one inch above the top of an ordinary table; then + electricity can be removed from the surface of the table immediately beneath the wax by means of a proof-plane, and tested with a +ly charged electroscope. Hold the excited wax near the wall, and positive electricity will then similarly be found upon the surface of the wall. These latter experiments should be repeated using excited glass instead of excited wax; the table, the wall, and any other body near which the excited glass is placed will then be found to be charged negatively by the inductive action of the electrified glass.

43. Cause of the Attraction of Neutral Bodies.—To understand the reason why a neutral body moves, or tries to move, towards an electrified body placed near it, we must bear in mind (1) the inductive action of the electrified body and (2) the law of inverse squares. Consider the state of the neutral insulated pith-ball, $a\,b$, when the excited glass rod, c, is brought near to it (Fig. 21). The electricity of the pith-ball is decomposed, positive being

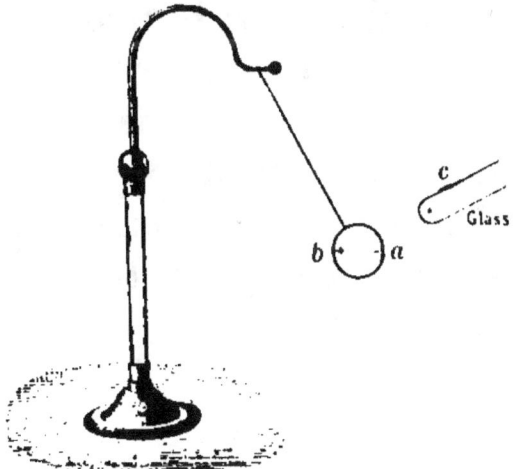

FIG. 21.—Attraction of a neutral body (glass rod and pith-ball).

repelled to the distant side, b, while the negative accumulates on the side, a, so that it may be as near as possible to the positive upon the glass rod. The pith-ball is now at once attracted and repelled by the excited glass. But the attraction is stronger than the repulsion, because the unlike electricities are *nearer to one another* than the like electricities. Let the distance, $c\ a$, be one inch, and the distance, $c\ b$, be two inches; then the force by which the ball is attracted will be *four times greater* than the force by which it is repelled. The pith-ball therefore moves towards the excited glass.

44. Neutral Bodies are more strongly attracted when uninsulated.—If the neutral body be *not* insulated, as in the case of an egg-shell resting upon a table, then its repelled electricity escapes into the earth, while electricity of the opposite kind to that upon the attracting body accumulates upon the side of the egg-shell nearest to it. As the egg-shell rolls after the excited glass rod, the negative electricity upon the egg-shell is continually changing its place, so as always to be upon that part of the egg-shell which is nearest to the positive electricity upon the excited glass. From this it is clear that an uninsulated neutral body will be more strongly attracted than an insulated one. In the former case there is attraction only; in the latter there is repulsion as well as attraction.

VIII.—THE ELECTROPHORUS.

45. Construction of the Electrophorus—46. Construction of the Collecting Plate—47 The Tea-tray Electrophorus—48. Experiments with the Electrophorus.

45. Construction of the Electrophorus.—The electrophorus is an electrical machine which depends, for its action, upon the principle of induction. The best form is shown in Fig. 22, where a is a circular thin sheet of vulcanite about twelve inches in diameter, known as the generating plate, which rests upon a brass plate, b, called the sole. Resting upon the vulcanite is shown the collecting plate, c, also made of brass, and provided with the insulating glass handle, d. Removing c, we proceed to (negatively) electrify the generating plate by briskly rubbing and flapping it with cat's fur or with a fox's brush. As we continue the rubbing, positive electricity is attracted on to the sole, and this reacts upon the generating plate, enabling us to get much more negative electricity upon a than we could if b were absent (Fig. 23).

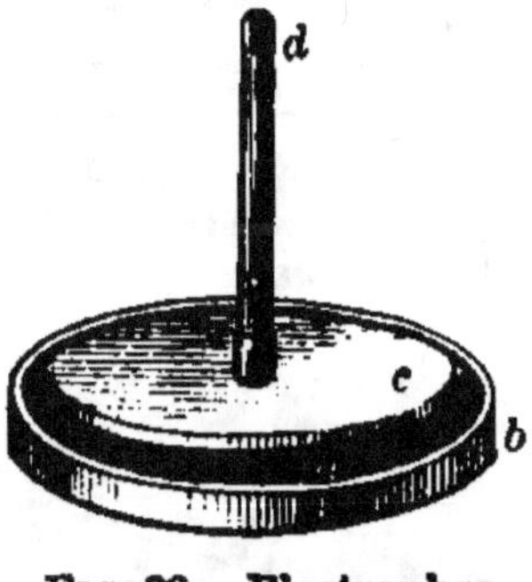

Fig. 22.—Electrophorus: a, generating plate; b, the sole; c, collecting plate; d, insulating handle.

Now take the collecting plate by the extremity

of its handle, and place it upon *a;* instantly its neutral electricity is decomposed (Fig. 24), + being

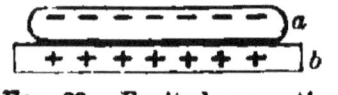

Fig. 23.—Excited generating plate with sole

attracted to the lower surface, while − is repelled to the upper. The two plates, *c* and *a*, in reality only touch at a few points, and as *a* is a non-conductor, the two electric fluids cannot unite, except at those points. Now touch the collecting plate with any conductor, as the finger (Fig. 25), and the repelled − electricity will escape to the earth, while more + electricity will stream upon *c*. Remove the finger, and lift the brass collecting plate by its handle, and it will be found to be positively electrified (Fig. 26). Approach the knuckle to its edge, and a bright electric spark will be obtained, due to the union of the + on *c* with − attracted by it to the extremity of the knuckle. The collecting plate may be replaced upon the generating plate, touched once more, and again raised, when it will be found fully charged as before; and this may be repeated for a considerable number of times. The instrument derives its name from *elektron*, electricity, and *phoreo*, I carry, because the charged collecting plate can be readily carried about from place to place.

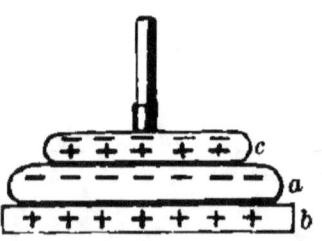

Fig. 24.—Inductive action of generating plate on collecting plate.

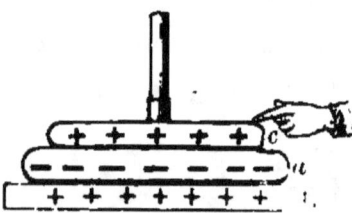

Fig. 25.—Escape of repelled − electricity from collecting plate.

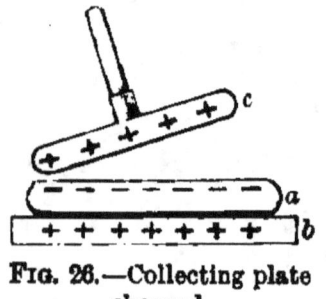

Fig. 26.—Collecting plate charged.

46. Construction of the Collecting Plate.—Instead of vulcanite, any non-conductor may be used for the collecting plate, as varnished glass, or a resinous cake made by melting together equal parts by weight of shellac, resin, and Venice turpentine. Sulphur or sealing-wax melted and run into shallow tin vessels, as the covers of canisters or biscuit-tins, answers well for small instruments, but these two substances are rather brittle. The electrophorus was invented by Professor Volta, an Italian, in the year 1774.

47. The Tea-tray Electrophorus.—Support a metal tea-tray upon four varnished glass tumblers. The tumblers used for this purpose, and indeed the glass employed in all electrical experiments, should be clear and of good quality; some of the very common glass now sold does not insulate at all well. The tumblers will insulate much better if varnished. This may be done by giving them, while warm, a thin coat of a varnish made by dissolving shellac in the best methylated spirit.

Take a piece of good, stout, brown paper, about the same size, or rather larger than the tray, make it hot in front of a clear fire, lay it upon a dry table, and brush it smartly with a common clothes-brush. Raise the paper by the two ends, and place it upon the insulated tray. The negative electricity produced by friction upon the paper attracts positive to the upper surface of the tray, and repels negative to the lower. Touch the tray, and the negative escapes, producing a spark. Now remove the paper, and the + electricity with which the tray is left charged will yield another good spark. In this arrangement the generating plate (the brown paper) is the one which is removed, while the collecting plate (the tray) remains stationary.

48. Experiments with the Electrophorus.—(1.) Charge the collecting plate (as in Fig. 26), and bring its edge just above a burner from which coal-gas is escaping. The gas will be ignited by the spark which will pass from the plate to the burner.

(2.) Place some light bodies, as bits of Dutch metal, upon the collecting plate, and set it down upon the excited generating plate. The light substances immediately rise into the air, for they become charged (like the upper surface of the plate on which they rest) with negative electricity, and being free to move they are repelled. Now touch the collecting plate, so as to allow this negative electricity to escape, and replace the light bodies upon it. Raise the plate by its insulating handle, and the light bodies again immediately fly from it, for they now receive a charge of the positive electricity which, when the plate is lifted, spreads all over its surfaces.

IX.—DISTRIBUTION OF ELECTRICITY.

49. Electricity resides upon the Surface of Conductors—50. Faraday's Net—51. Biot's Hemispheres—52. Charge depends upon Area of Surface—53. Limit of Charge—54. Density of a Surface Charge—55. Surface Distribution—56. Escape of Electricity from Points: the Electric Wind—57. Electric Mill.

49. Electricity resides upon the Surface of Conductors.—Let the hollow brass ball, A (Fig. 27), having a hole in its top, and insulated by the glass tube, B, be electrified as strongly as possible. If its *outer* surface be now touched with a proof-plane, clear signs of electricity will be obtained; but when the proof-plane is passed through the hole in the ball and made to touch its *inner* surface, no electricity will be found there. The experiment

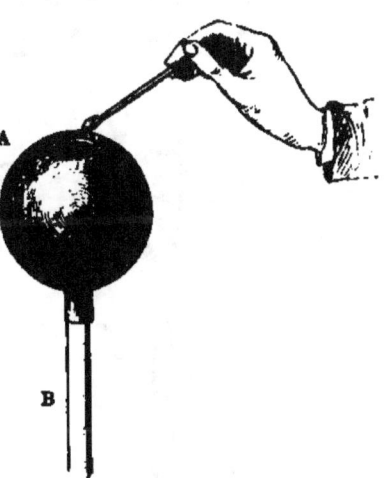

FIG. 27.—Electricity on surface only.

may be repeated with various hollow, insulated, electrified conductors, as a hat, a sauce-pan, pewter-pot, etc.; and in every case the electricity will be found upon the *outside* only. For this reason, a solid ball, weighing many pounds, cannot be charged more strongly than a hollow ball of the same diameter, weighing only a few ounces.

This is what we should expect, since the electric fluid is *self-repulsive;* each part of the same fluid with which any body is charged tries to get as far from every other part of that fluid as possible, and this can only be the case when the electricity is all upon the exterior surface of the body. Moreover, the electricity on an insulated conductor acts inductively on all surrounding bodies, such as the surface of the table on which it stands and the walls of the room. The electricity so attracted (which will be of the opposite name or kind) then reacts upon the insulated conductor, attracting and so drawing the electricity with which it is charged to its outer surface.

50. Faraday's Net.—In Fig. 28 is shown an insu-

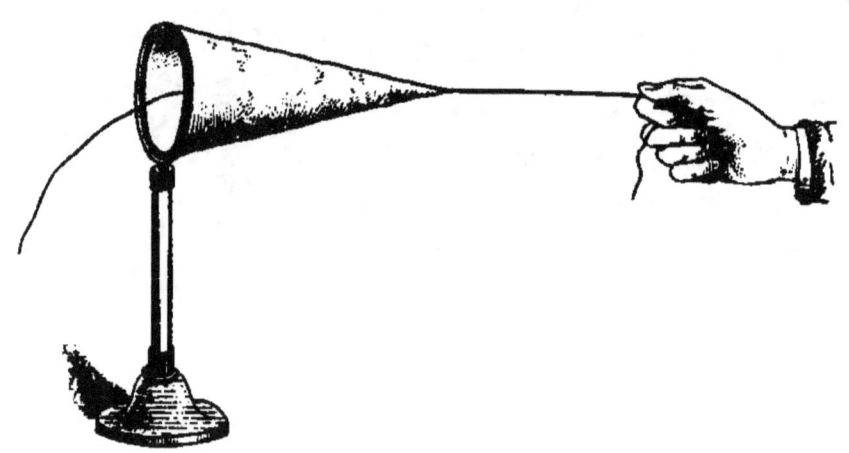

FIG. 28.—Faraday's Net.

lated conical net, made of cotton, to the apex of which a silk thread is attached. Let the net be electrified while in the position shown in the figure. With a proof-plane, electricity can be removed from what is then its *outer* surface; but none will be found inside the net. Now, by means of the silk thread, turn the net inside out, when it will be found that the electricity has also changed sides, so as to be still on the *outer* side of the net; and the same thing will happen if the net is turned inside out again and again—the electric fluid will always pass to the outside of the net.

51. Biot's Hemispheres.—Let the insulated metal ball, A (Fig. 29), be charged with electricity. Now take the two brass hemispheres, B and C, made to fit the ball and provided with insulating handles, and place them so as to cover the ball, A. On re-

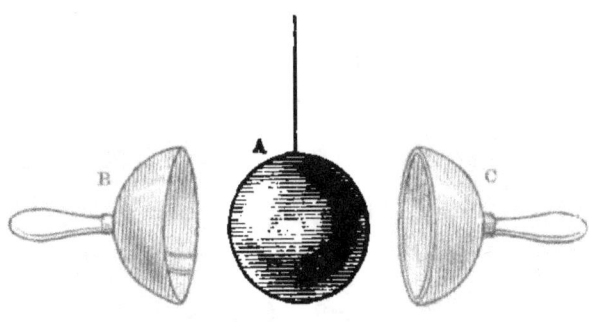

FIG. 29.—Biot's Hemispheres.

moving the hemispheres they will each be found to be charged, while A will have lost its electricity, which passed outward to the surfaces of B and C, when these were made to cover it. This experiment was devised by a French electrician named Biot.

52. Charge depends upon Area of Surface.—When an electrified *non-conductor* is touched by the finger, only the electricity of the point touched is discharged or neutralized. The electricity on the adjoining portions of the non-conductor is unable to flow over the surface, and cannot, therefore, take part in the action. But when an insulated electrified *conductor* is touched by the finger, or is put in any way in connection with the earth, then it is at once discharged entirely, because the electricity can flow over the whole surface to the point which is touched.

53. Limit of Charge.—For every insulated conductor there is a point beyond which it cannot be charged, or electrified. If we try to electrify it beyond this point we fail, because the electricity escapes into the air and along the insulating support as rapidly as we add it to the body. It is like pouring water into a vessel which is already full. This *capacity* of a body to receive electricity depends upon the area of its surface; the greater the size of the body, the greater is its capacity. We may compare this to the measures used for liquids; a quart measure can hold twice as much as a pint, because it is larger. In just the same way the capacity for electricity of a little ball (a leaden bullet, for instance) is very small, while the capacity of a cannon-ball will be just as much greater as its radius is greater than that of the bullet. If the radius of the bullet $= \frac{1}{4}$ inch, and that of the cannon-ball $= 5$ inches, then the electric capacity of the latter will be $5 \div \frac{1}{4} = 20$ times greater than that of the former. It is plain that if the cannon-ball and the bullet were each to be fully charged with electricity, we should get a *discharge* from the ball twenty times greater than from the bullet.

When a charged body is made to touch an uncharged body of the same size and shape (both being insulated) the charge of electricity is equally divided between them.

54. Density of a Surface Charge.—By electric density we mean the amount of electricity accumulated on any given area of the surface of a body. If we compare the electric fluid to a little ocean surrounding a body, then the density of the electricity is comparable to the *depth* of the fluid forming the ocean.

When a charge of electricity is imparted to a *small* insulated conductor, the *density* is clearly much greater than if the same charge were imparted to a larger body; just as water rises to a greater *height* in an ordinary pint measure than in a quart measure, when the same quantity is poured into each. For this reason, the larger a conductor the greater is its capacity for electricity, and the greater is the charge which it can receive, and the discharge which it will afford after it has been fully charged.

55. Surface Distribution.—Upon a *sphere* (Fig. 30)

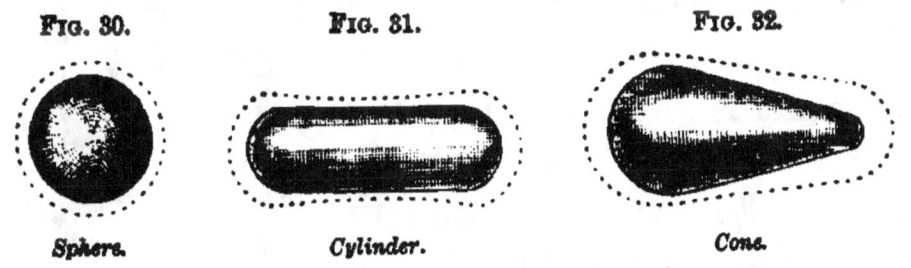

Surface Distribution of Electricity.

it is evident that the electricity will be *equally* distributed all over the surface: with the proof-plane we find that we remove an equal quantity of electricity (as shown by an electroscope), no matter what part of the surface of the sphere is touched. But on a *cylinder* (Fig. 31) the case is different; the density of the charge is found to be much greater at the ends than in the middle. So, also, upon a *cone* there is a great accumulation of the electric fluid at the apex (Fig. 32). Speaking generally, we may say that the density increases with the curvature—the sharper the curve at any point of the surface, the greater the density.

56. Escape of Electricity from Points: the Electric Wind.—From the above paragraph (55) it is plain that if there is a sharp point upon the surface of a charged conductor, the density of the electricity upon such a point must be very great. The result is that the molecules of the air touching the point become electrified (Fig. 33), and as soon as this happens they are *repelled*, thus producing a current of air (the so-called electric wind), which moves from the point, carrying away electricity, and thus discharging the body upon which the point is situated. This electric wind may be felt with the hand, and it will blow aside the flame of a candle (Fig. 34).

FIG. 33.—Escape from Points.

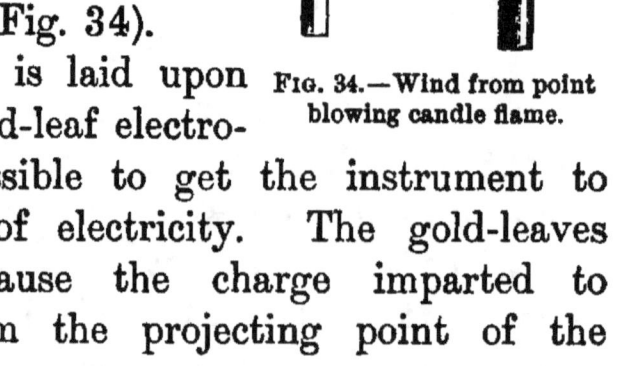

FIG. 34.—Wind from point blowing candle flame.

When a needle is laid upon the plate of a gold-leaf electroscope, it is impossible to get the instrument to retain a charge of electricity. The gold-leaves fall together because the charge imparted to them escapes from the projecting point of the needle.

57. Electric Mill.—If four bent wires are arranged so as to form a cross, which is then balanced at its centre upon an insulated pivot (Fig. 35), and strongly electrified, the repulsion of the electrified air at the four points will cause the wires to move round upon the pivot in the *opposite* direction to that in which the air is driven away; just as when we give any object a violent push with the hand, we are ourselves pushed backward.

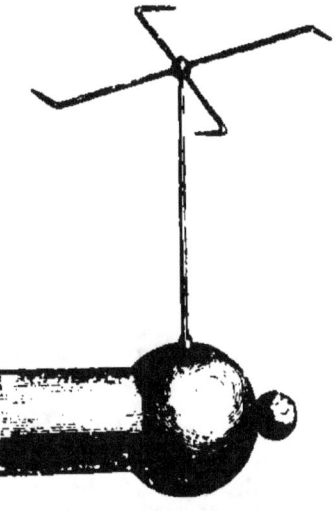

Fig. 35.—Electric Mill.

X.—ELECTRICAL MACHINES.

58. Contrivances for obtaining large Quantities of Electricity—59. The Cylinder Electrical Machine—60. The Plate Electrical Machine—61. How to manage Electrical Machines—62. Power of Electrical Machines—63. The Rubber supplies Negative Electricity—64. Causes which influence the Working of Electrical Machines.

58. Contrivances for obtaining large Quantities of Electricity.—The men who conducted the first scientific experiments on frictional electricity, more than two centuries ago, soon became dissatisfied with the small amount of electricity which they obtained by rubbing glass tubes with silk. They successfully endeavoured to construct instruments which should yield larger quantities of the remarkable force whose properties they were so earnestly engaged in studying. The first electrical *machine* was a globe of sulphur, turned by a handle fastened to an axis passing through the globe, and rubbed by the dry hand. Then a glass globe was substituted for that of sulphur, and the hand was replaced, as a rubber, by a cushion covered with amalgamed silk. The part of the machine called the *prime conductor* was added by a German named Boze, in 1741.

59. The Cylinder Electrical Machine.—A common form of machine in use at the present day is represented in Fig. 36. Here we see a glass cylinder, supported at each end by glass uprights, and turned by a handle which is fastened to one end of an axis

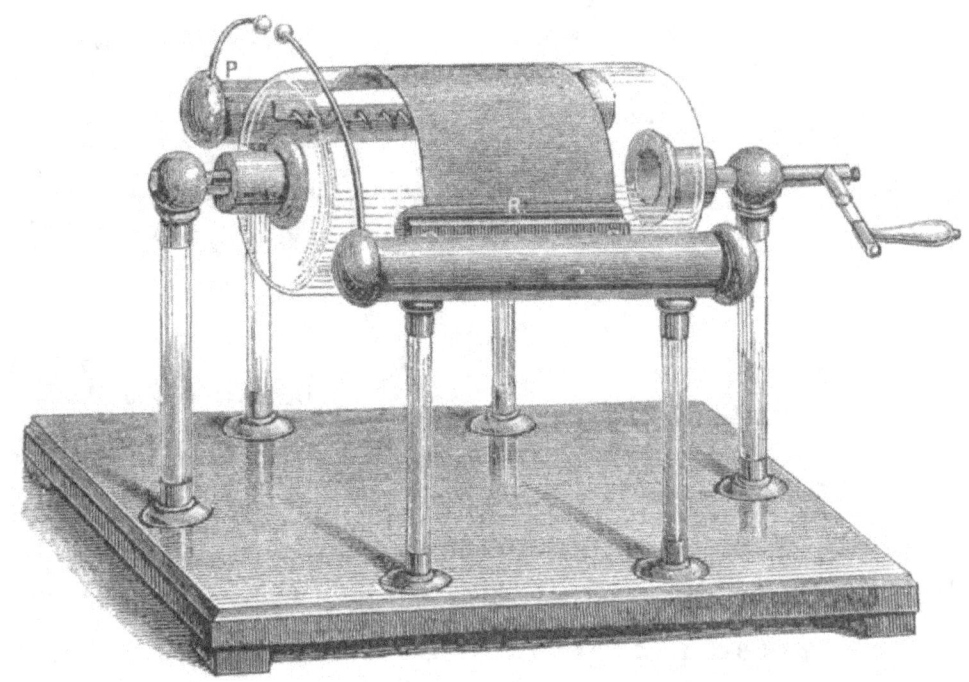

Fig. 86.—Cylinder Machine: R, Rubber; P, Prime Conductor.

that forms part of the cylinder. The rubber, R, consists of a leather cushion stuffed with hair and covered with amalgamed silk. A flap of oiled silk extends from the rubber over the top of the cylinder, in order to prevent the escape of electricity into the air. The prime conductor, P, consists of a metal cylinder, supported by a varnished glass rod, and provided with a row of pin-points on the side facing the glass cylinder.

When the handle is turned + electricity is produced upon the glass by its friction against the silk rubber. When this positively electrified glass comes opposite to the points of the prime conductor it attracts negative electricity, which streams through the points, leaving the prime conductor charged with positive electricity (Fig. 37).

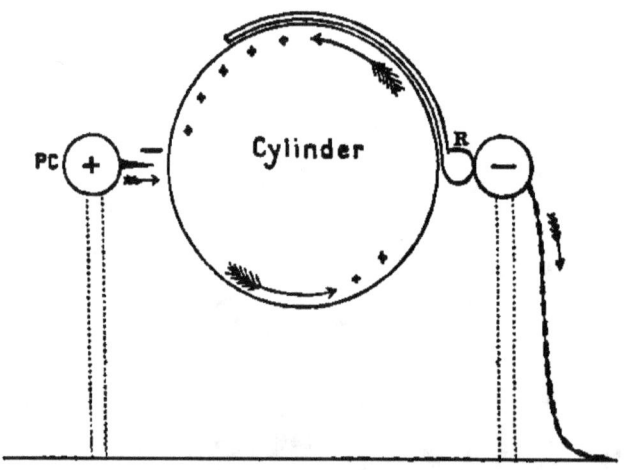

Fig. 37.—Section of Cylinder Machine: P C, Prime Conductor; R, Rubber.

The rubber must be connected with the earth by means of a metal chain or wire passing from the cushion to the table or floor, or, better, fastened to the gas-pipe. With a small machine it is sufficient to place the hand upon the rubber, pressing it against the glass; the − electricity then escapes through the body to the ground. For a very cheap, home-made cylinder machine a common glass wine-bottle may be used, though the clear glass jars used by confectioners answer much better; but a glass cylinder made for the purpose, and provided with a hole at each end through which to pass the axle, may be bought for about eighteenpence.

60. The Plate Electrical Machine.—Instead of a cylinder, we may use a flat plate or disc of glass or vulcanite, usually from twelve to eighteen inches in diameter (Fig. 38). Two rubbers, $a\,d$ and $b\,d$, are then employed, each clasping the glass so as to rub both sides of the disc, and each provided with a silk flap. The prime conductor consists of the insulated and connected pieces of brass tubing, $e\,e$, provided with points on each side to discharge the negative electricity to the glass.

61. How to manage Electrical Machines.—Every part of the machine must be warm and *dry*, or it

Fig. 38.—Plate Machine.

will not work well. Before using it, the rubbers should be taken off and, with the rest of the machine, placed before a clear fire to dry. While standing there, the machine must be watched and the glass cylinder or plate turned round occasionally, so that it may not crack by being unequally heated. Fresh amalgam should occasionally be spread upon the rubbers, the old stuff having been previously scraped off. The insulating support of the prime conductor must be frequently rubbed with hot dry flannel, especially if the weather is damp. While the machine is being worked, it is a great advantage to clamp it firmly to the table, so that it cannot shift about. If there is no clamp, one person should hold the machine steady by placing his hands upon its base, while a second person turns the handle with one hand and (if it is a cylinder machine) presses the rubber against the glass with the other.

62. Power of Electrical Machines.—From the prime conductor of an ordinary electrical machine we ought to obtain abundant supplies of positive electricity. A machine with a glass *cylinder* about ten or twelve inches long by six to nine inches in width should, in dry weather, give a rapid succession of sparks of from one to two inches in length; a machine with a *plate* eighteen inches in diameter ought to give sparks from two to four inches in length.

63. The Rubber supplies Negative Electricity.—When a supply of negative electricity is desired, it is convenient to use a cylinder machine whose rubber is attached to a brass cylinder supported by legs or pillars of glass or vulcanite (Fig. 36). We must then connect the prime conductor with the earth by means of a chain, or by placing one hand upon it. When the machine is worked, sparks of negative electricity can then be obtained from the metal surface to which the rubber is attached.

64. Causes which influence the Working of Electrical Machines.—In all these machines the use of sharp corners or points must be avoided, with the exception of the row of points by which negative electricity passes from the prime conductor to the excited glass. It is our object to accumulate as great a quantity of electricity as possible upon the prime conductor; so that that important part of the machine should be well polished, well insulated, as large as possible, and supplied with rounded ends. If a needle is placed upon the prime conductor, it will be found impossible to obtain a spark of any length, for electricity escapes from the point almost as rapidly as it is produced by the action of the machine. A plate machine yields a *longer* spark than a cylinder machine of equal size, because the distance between the rubber and the prime conductor is greater. If a cylinder machine be worked in the dark, frequent flashes of light, each marking a combination of the two electric fluids, will be seen to pass between the prime conductor and the rubber, darting over the surface of the glass.

XI.—ELECTRICAL MACHINES.
(*Continued.*)

65. Experiments with the Electrical Machine—66. Electric Sparks from the Human Body—67. The Electric Spark—68. Head of Hair electrified—69. Electrical Hail—70. Quadrant Electroscope—71. Historical Note.

65. Experiments with the Electrical Machine.—Any experiments which were not very successful when performed with excited rods of glass or sealing-wax should be repeated with the aid of the much larger store of electricity furnished by the machines we have now described.

66. Electric Sparks from the Human Body.—To prove that our own bodies may be electrified, we must have the means of insulating them. For this purpose it is best to stand upon an *insulating* stool, which consists of a wooden board supported by four varnished glass legs: glass tumblers will furnish a support for the board if the ordinary glass supports cannot be obtained. Standing on this stool, let the experimenter place his hand upon the prime conductor of a machine in action. While in this position, the person becomes, as it were, a part of the prime conductor, and negative electricity is withdrawn from his body by the action of the machine, leaving him charged positively. If his hair is dry and tolerably short, it will then stand up "like quills upon the fretful porcupine," while sparks may be drawn from his knuckle, ear, nose, or, in fact, from any part of his body. If the knuckle of the electrified person be placed just over a jet from which coal-gas is escaping, the spark which will pass will light the gas.

67. The Electric Spark.—We know that a spark of light, accompanied by a crackling or snapping noise, is the result of the sudden combination of the two opposite kinds of electricity. To notice the exact appearance of the electric spark, the experiments should be conducted in the dark.

When any conductor, as the knuckle, or a brass ball connected with the earth, is brought near to a charged prime conductor, the + electricity on the latter acts inductively upon the conductor, attracting negative to the part nearest the prime conductor, and repelling the positive to the ground. When the ball is brought very near—within, say, half-an-inch—the two electricities unite, causing the air between them to become white-hot, and thus producing a short, thick, bright flash or spark (Fig. 39).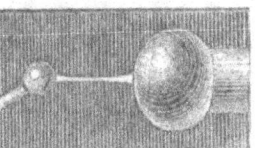

Fig. 39.—Electric spark.

Now remove the brass ball to a distance of two or three inches; the spark is then irregular in shape and branched (Fig. 40), probably because the electricity seeks to pass in the easiest way from the one point to the other, and avails itself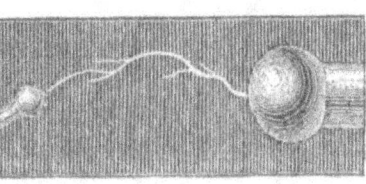

Fig. 40.—Branched spark.

of the dust particles in the air for this purpose.

With a powerful electrical machine sparks can be obtained at a distance of from six to twelve inches,

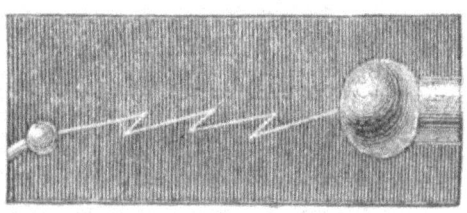

Fig. 41.—Forked spark.

or even more. These long sparks take a zigzag path, reminding one strongly of forked lightning (Fig. 41). Lastly, if the distance be too great for any sparks to pass, we have what is called the *brush discharge*, visible in the dark as a fan-

Fig. 42.—Brush discharge.

shaped streamer of violet light escaping from the prime conductor (Fig. 42). By the use of a needle on the P C, the brush discharge may be obtained from even a small machine.

68. Head of Hair electrified.— It is an amusing experiment to mount the head of a wooden doll (which should have rather long, dry,

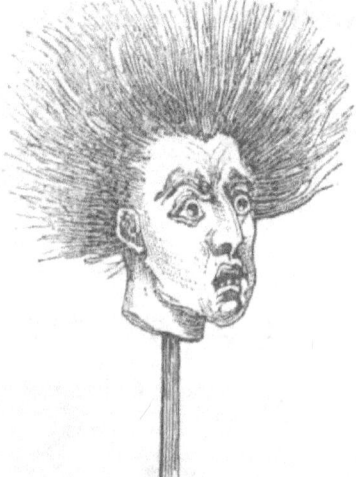

Fig. 43.—Head of hair electrified.

and well-combed hair) upon a short piece of brass rod, which may then be fitted into a hole in the prime conductor. When the machine is worked, every hair is electrified, and as "like electricities repel," the hairs rise up and stand out in a very remarkable manner (Fig. 43).

69. Electrical Hail.—Fig. 44 represents a glass bell-jar, standing on a wooden base covered with tinfoil, and having a brass knob instead of a stopper. A dozen pith-balls are placed within the jar, lying upon the tinfoil. Connect the knob with a charged prime conductor by means of a copper wire. Immediately the pith-balls rush upwards to the knob, but no sooner do they touch it than they are violently repelled. In this manner they continue to rush up and down as long as the machine to which the prime conductor is attached is worked.

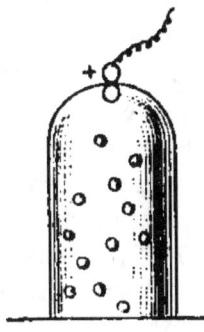

FIG. 44.—Electric Hail (pith-balls under bell-jar).

The knob of the jar is charged positively, and therefore attracts the pith-balls; but when they touch it, they themselves receive a like charge, and are therefore repelled. On touching the tinfoil the balls lose their electricity, and are then again attracted by the knob.

70. Quadrant Electroscope.—The piece of apparatus shown in Fig. 45 is used to indicate the degree to which a prime conductor, or indeed any electrified body, is charged. It consists of a brass rod, A B, by the side of which hangs a straw, or light piece of wood, bearing a pith-ball, E, at the end. A scale or quadrant, divided into degrees, is fixed to the side of the rod. When this quadrant electroscope is placed upon the uncharged prime conductor, the straw hangs parallel to the rod; but when the machine is worked, the pith-ball and the brass rod are charged with the same kind of electricity, and they therefore repel one another. The former, being free to move, rises, and

FIG. 45.—Quadrant Electroscope.

the angle through which it moves indicates roughly the amount of the charge of electricity.

71. Historical Note.—First electrical machine made by Otto von Guericke of Magdeburg in 1675. Hauksbee, in 1705, substituted a globe of glass for the ball of sulphur; Boze of Wittenberg added the prime conductor in 1741; Gordon and Nairne used a glass cylinder (instead of a globe) about 1750; and in 1760, Planta introduced circular glass discs in what was called the plate machine, which was improved by Ramsden (1768), Van Marum, Cuthbertson, Harris, and others. In 1862, Winter of Vienna added to the prime conductor a stout wire, bent into a circle, about three feet in diameter, and enclosed in a wood casing (Winter's ring), by which the length of spark was greatly increased. The largest plate machine ever made was that formerly in use at the Polytechnic, in London, which had a glass disc 7 feet in diameter. It was turned by steam power, and gave sparks of from 10 to 15 inches in length. In 1840, Lord Armstrong constructed a hydro-electric machine, consisting of an insulated steam boiler, which was electrified by the friction of the steam escaping through small openings: sparks 22 inches long were obtained. In 1865, Holtz of Berlin invented a machine, which may be called a "continuous electrophorus" (see Fig. 46). The electricity is produced by the simple rotation of a glass plate in the neighbourhood of a fixed (charged) plate, without any friction against rubbers.

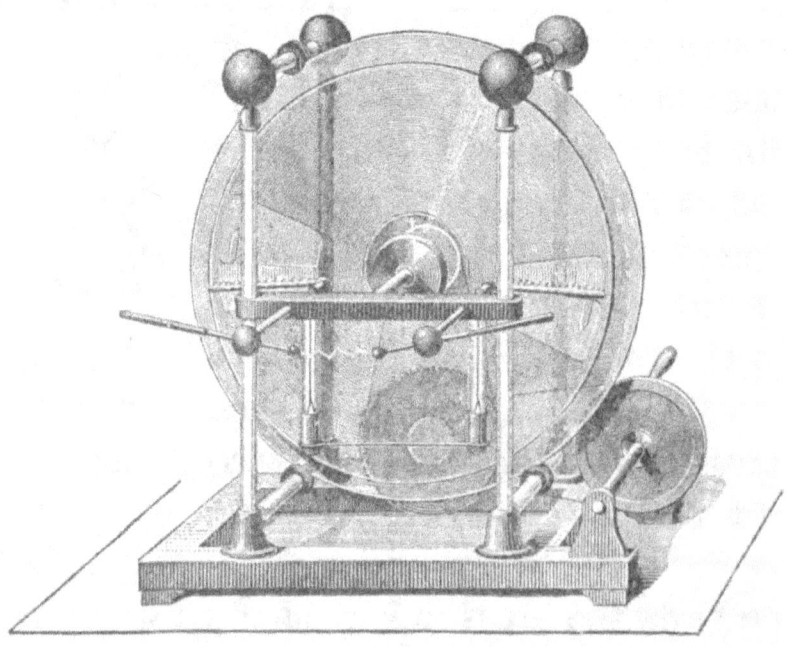

FIG. 46.—Holtz's Electrical Machine.

XII.—ELECTRICAL CONDENSERS.

72. Use of Condensers—73. The Coated Pane, or Bevis's Plate—74. Discharging Tongs—75. The Condenser of Epinus—76. Experiments with the Condenser—77. Discovery of the Condenser—78. Principle of the Leyden Phial.

72. Use of Condensers.—We have seen (par. 53) that there is a limit to the degree to which we can charge an insulated conductor, under ordinary circumstances. Now we know that there are certain substances, called dielectrics, or insulators, which will not permit electricity itself to pass, but which will allow induction to take place through them. Thus an excited glass rod will attract a negatively-charged pith-ball, although a sheet of glass is placed between the two bodies (Fig. 47).

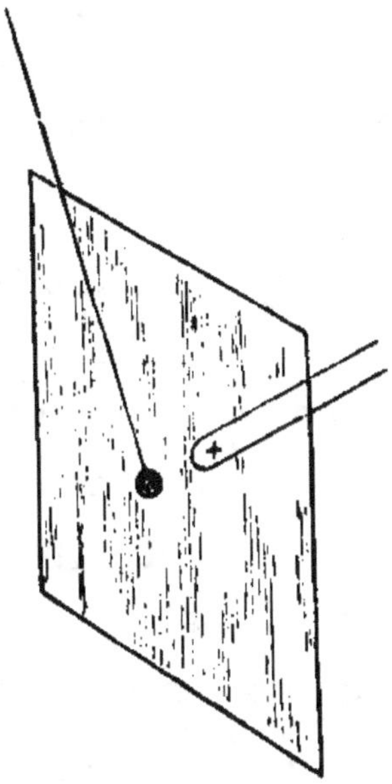

FIG. 47.—Pith-ball attracted through glass.

73. The Coated Pane, or Bevis's Plate.—Let a smooth sheet of tinfoil, about eight inches square, be laid upon a piece of varnished glass ten inches square. Connect the tinfoil by a wire with the prime conductor of an electrical machine. Although charged by this means to its fullest capacity, the tinfoil will yield but a very small spark on being afterwards discharged. The fact is, that the *capacity* of such a piece of tinfoil is very small. But may it not be possible to *increase* the capacity of the tinfoil, so as to enable it to receive a much greater charge of electricity? For this purpose, let a second piece of tinfoil be placed *underneath* the varnished glass, and connected with the earth by a thin wire. When the upper sheet of foil is electrified positively, the electricity acts inductively through the glass, attracting negative electricity from the earth to the lower piece of foil. This negative electricity then *reacts* upon the positive fluid on the upper foil, attracting it and binding it to the metal plate, as it were; and we find that we are thus able to accumulate or condense upon these two pieces of tinfoil many times the quantity of electricity which either of them could hold separately. For this reason such a piece of apparatus is known as a *condenser*. If the two pieces of charged tinfoil are now connected by any conductor, a bright spark will be seen and a considerable shock will be felt if the discharge takes place through the human body. To secure the best results another condenser should be made, having the sheets of tinfoil carefully pasted upon the opposite sides of the pane of glass. The apparatus is then known as Bevis's plate.

74. Discharging Tongs.—To allow the electricity accumulated upon one plate of a condenser to unite with that of the opposite kind which is found upon the other plate, we may make use of the instrument called the discharging tongs, or simply the dischargers (Fig. 48). This consists of a curved piece of thick brass wire, hinged at the centre, having a brass knob at each end, and provided with an insulating handle or handles. To use it, we should first cause one knob to touch the plate connected with the earth, and then bring the other knob to touch the insulated plate. Since the brass wire is a good conductor, it offers a path for the electricities to combine, the result being the production of a bright spark, accompanied by a smart crack.

FIG. 48.—Discharging Tongs.

75. The Condenser of Epinus.—Another form of condenser is shown in Fig. 49. It consists of two

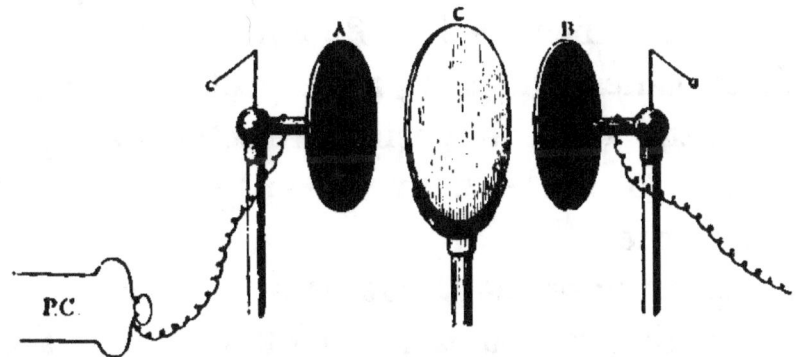

FIG. 49.—Epinus's Condenser.

insulated metal discs, A and B, separated by a disc of vulcanite, C. The plate A is connected with the prime conductor of an electrical machine, while B is connected with the earth. The positive electricity which A receives from the machine attracts negative from the earth on to B. This negative in turn reacts on the positive upon A, and *holds it there*. Thus we can accumulate much more electricity upon A and upon B when they are close

70 ELECTRICAL CONDENSERS.

to one another, than we can when they are far apart.

76. Experiments with the Condenser.—If only the plate B is touched with the finger, no shock is felt, no discharge takes place, because when B alone is touched no path is offered for its electricity to combine with that upon the insulated plate A. When B is held with one hand, while A is touched with the other, a smart shock is felt, for the discharge then takes place through the human body. When the plate A alone is touched, a shock is felt, for A is then connected with the earth by means of the body of the person touching that plate; and B, we know, is connected with the earth also. Let a boy keep one finger upon B, while a second boy touches the plate A; then each boy will receive a shock, for their bodies, and the ground between their feet, offer a path for the positive electricity upon the one plate to unite with the negative electricity upon the other.

77. Discovery of the Condenser.—The town of Leyden is situated in Holland; it is the seat of a great university. In the year 1746, two students of electricity, named Cuneus and Musschenbroek, who lived at Leyden, were endeavouring to electrify some water contained in a glass phial or bottle, by allowing sparks to pass from a prime conductor to the head of a long nail, which passed through a cork, and dipped into the water, of which the bottle was about half-full (Fig. 50). Holding the bottle in

FIG. 50.—First Leyden Jar.

one hand, Musschenbroek allowed, as he thought, a sufficient number of sparks to pass to the nail, which he then attempted to remove with his other hand. On touching the nail he received a shock which caused him to drop the bottle, and which so terrified him that, in describing the experiment in a letter to a friend, he wrote that he would not receive such another, "no, not for the kingdom of France!" The news of this wonderful experiment soon spread all over Europe. The name of the "Leyden jar" was applied to Musschenbroek's phial, and it was improved upon until it at last assumed its present well-known form.

78. Principle of the Leyden Phial.—In Musschenbroek's bottle it is not difficult to recognize the two plates, made of a conducting material and separated by a dielectric, which we know to constitute a condenser. The glass bottle is, of course, the dielectric; the water answers to the insulated plate or sheet of tinfoil; while the hand of the experimenter, clasping the outside of the bottle, furnishes an outer plate or coating. When Musschenbroek attempted to pull out the nail, he offered, as he found to his cost, a passage through his body for the two electricities to unite. It is true that that shock was probably not a severe one, but we must remember that it was totally *unexpected*.

XIII.—THE LEYDEN JAR.

79. Modern Form of the Leyden Jar—80. Construction of the Leyden Jar—81. How to charge a Leyden Jar—82. How to discharge a Leyden Jar—83. An Insulated Jar cannot be charged—84. Leyden Jar with Movable Coatings—85. The Residual Charge—86. Battery of Leyden Jars—87. Cascade Arrangement of Leyden Jars.

79. Modern Form of the Leyden Jar.—The form of Leyden jar now in common use consists of a glass bottle or jar, coated within and without with tinfoil, which must not come within an inch (or two inches if the jar is a large one) of the edge of the jar (Fig. 51). A stout brass wire with a metal knob at the end is fastened to the inner coating, and rises in the centre of the jar to a height of two or three inches above the edge.

80. Construction of the Leyden Jar.—To make a Leyden jar, we may select a large smooth tumbler of thin good glass, and paste tinfoil upon the inside and the outside, to within one inch of the edge (not forgetting to cover the bottom); the foil must be rubbed down with the thumb-nail until it lies smoothly, without any creases. Now cut a long piece of stout copper wire, and fasten a knob of metal, or of wood covered with tinfoil, to one end; bend the other end into a circle, so that it will stand inside the tumbler, upon the tinfoil which

THE LEYDEN JAR.

covers the bottom; fasten it down by dropping melted sealing-wax upon it, and the jar is complete (Fig. 51). This is a Leyden jar with an open top; but jars are also used whose tops are closed by wood or cork, through which the central wire passes. In that case the wire is usually connected with the inner coating by a piece of brass chain (Fig. 52). The exposed glass edge of the jar should, if possible, receive a thin coating of shellac varnish.

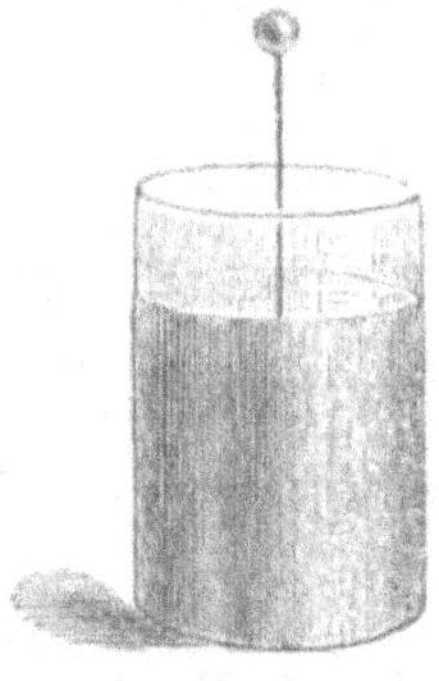

FIG. 51.—Leyden jar with open top.

81. How to charge a Leyden Jar.—Holding the jar by its outer coating, present the knob to a

FIG. 52.—Leyden jar with closed top.

charged prime conductor, and notice the electric sparks that pass from the knob. After a time the sparks cease, and this is a sign that the jar is fully charged. The + electricity upon the prime conductor attracted − electricity from the inner coating; and this negative streams out of the knob, leaving the inner coating of the jar charged positively. Then this positive electricity acts inductively through the dielectric—the glass—repelling the positive which is upon the outer coating (and which escapes to the earth through the body of the person holding the jar) and attracting negative. Finally, when the jar is "charged," its inner coating is positively and its outer coating negatively electrified.

If we desire to have positive electricity upon the

outer coating, it is only necessary to hold the jar by its knob, and so to present the *outer* coating to the prime conductor of an electrical machine. Of course it is possible to charge a Leyden jar by means of an electrophorus, or even with an excited tube of glass or of vulcanite; but a cylinder or a plate machine will charge a Leyden jar most rapidly and effectively. The glass of a Leyden jar is strained or squeezed by the pressure of the two layers of electricity, one upon each side. If the glass be very thin, it may even by this means be cracked, or perforated, when the jar is strongly charged.

82. How to discharge a Leyden Jar.—To discharge a Leyden jar, it is only necessary to allow the opposite electricities (the positive upon the inner and the negative upon the outer coating) to unite; and this must be done by connecting the two coatings by means of a conductor (Fig. 53). Press one knob of the discharging tongs against the outer tinfoil, and then bring the other knob to touch the knob of the jar. Instantly the two electricities combine, producing a dense and dazzling spark and a loud cracking sound. The hands may be used instead of the tongs; but the discharge then takes place through the body, producing a convulsive twitching of the muscles of the arm, and the sensation is, to most people, far from agreeable. When a charged Leyden jar is insulated, as by placing it upon a sheet of vulcan-

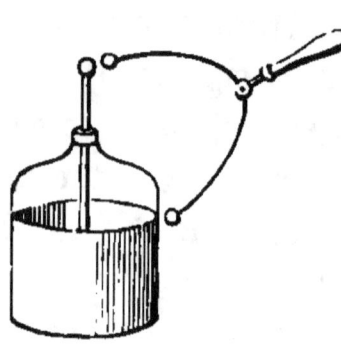

Fig. 53.—Discharging Leyden jar.

ite, we may touch either coating (one at a time) without fear of discharging the jar. With an uninsulated charged jar (standing, say, on an ordinary table) we may touch the outer coating; but if the finger be placed upon the knob connected with the inner coating, the discharge will take place through the body of the person and the ground.

83. An Insulated Jar cannot be charged.—If a Leyden jar be placed upon an insulating stool or table, or held by means of a piece of india-rubber, it will be found impossible to charge it properly. The reason is that the repelled positive electricity upon the outer coating cannot then escape to the earth. Neither is there then any path for more negative electricity to pass from the earth to become condensed upon the outer coating.

To show that electricity is *repelled* from the outer coating when a Leyden jar is being charged, it is only necessary to connect the outer coating of an insulated jar, by means of a wire, with a gold-leaf electroscope. As soon as sparks pass from the knob to the prime conductor, the gold leaves diverge, showing that electricity has been driven into them. On removing the wire by means of an insulator, and bringing excited glass near the electroscope, we shall obtain *increased* divergence; this proves that the repelled electricity was *positive*. When the knob of an uncharged jar is touched by the knob of a charged jar of the same size, the electricity divides itself equally between the two jars, each now having half the original charge.

84. Leyden Jar with Movable Coatings.—In a

charged jar the electricity is believed to reside upon the surfaces of the glass, and not upon the metal with which it is coated. To prove this, we use a jar with movable coatings, which may be made of sheet-tin (Fig. 54). Having charged such

Complete jar.   Glass jar.   Outer coating.   Inner coating.

Fig. 54.—Leyden jar with movable coatings.

a jar, we can take it to pieces by lifting out first the inner coating by means of a glass rod, and then removing the glass jar. Touch both the coatings so as to make sure that they contain no free electricity, and then put the jar together again in the same way as it was taken to pieces. On then connecting the outer and inner coatings by means of the dischargers, a bright spark will be obtained, proving that when the coatings were removed the electricity was left behind upon the surfaces of the glass.

85. **The Residual Charge.**—It would appear that a small part of the electricity upon each surface of the glass penetrates into its pores, soaks into the glass as it were, in its endeavour to unite with the opposite kind of electricity which is situated upon the other surface. Another reason for this action would be the self-repulsion of each kind of electric-

ity. Consequently we find that after a large and strongly-charged Leyden jar has been discharged, and then allowed to stand for five minutes, that we can obtain a *residual* or *secondary* discharge by again connecting the two coatings. During the interval, the electricity on each surface has oozed out of the pores of the glass, and is therefore able to produce this residual discharge. For this reason it is necessary to be careful in handling large jars which have just recently been discharged.

86. Battery of Leyden Jars.—The ordinary Leyden jars are made either as "pints" or "quarts." Sometimes glass jars having a capacity of one or two gallons are used, but they are inconveniently large and clumsy. When it is desired to obtain a more powerful discharge than can be produced by a single quart jar, it is better to unite several such jars into what is called a Leyden battery than to endeavour to make a single very large jar. All that is required in order to enable a number of small jars to act like one large jar is to connect all their inner coatings by one conductor, and all their outer coatings by another conductor. This is most conveniently effected by placing all the jars within a shallow box lined with tinfoil (by which all their outer coatings are connected), and connecting all their knobs by pieces of brass tubing or wire (Fig. 55).

Fig. 55.—Battery of Leyden jars.

The tinfoil lining of the box must be connected with the earth by means of a metallic chain.

87. Cascade Arrangement of Leyden Jars.—Franklin proposed to charge economically a series of Leyden jars by insulating all except the last jar, and then connecting the outer coating of the first jar with the inner coating of the second, and so on (Fig. 56). By this means the electricity re-

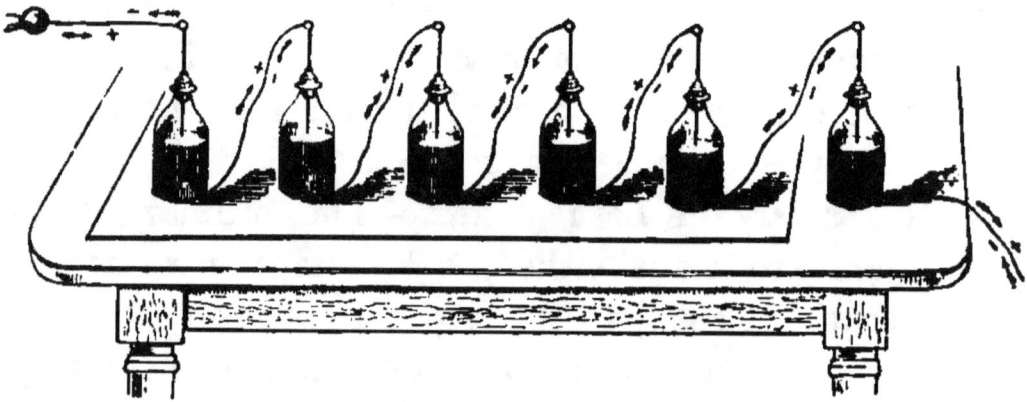

Fig. 56.—Charging Leyden jars by cascade.

pelled from the outer coating of the first jar would flow into and charge the second jar; and in like manner the electricity would flow in an imaginary *cascade*, from jar to jar, when the first one only was directly charged. There is no real advantage, however, in this arrangement, for the charge thus accumulated upon such a series of jars is found to be no greater than could be obtained upon any *one* of them separately.

XIV.—EFFECTS OF ELECTRICITY.

88. Luminous Effects—89. The Spiral Tube—90. Chemical Effects—91. Heat Effects—92. Mechanical Effects—93. Magnetic Effect—94. Physiological Effects.

88. Luminous Effects.—The appearance of the electric spark in ordinary air has already been described. If the density of the air be diminished, the length of the spark will be increased up to a certain point. Let $a\,b$ (Fig. 57) be a glass tube having a short piece of platinum wire passed through each end. By means of an air-pump let nine-tenths of the air within the tube be removed through the opening at c.

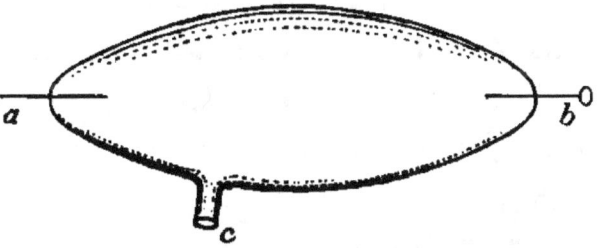

FIG. 57.—Vacuum Tube. a must be connected with some source of electricity, and b with the earth.

It will now be found that the spark from an electrophorus, or an electric machine, which in ordinary air could only travel across a space of, say, one inch, can pass with ease across the nine or ten inches of rarefied air which separate a from b. Tubes containing some such gas as hydrogen, nitrogen, or carbonic acid gas may also be used, and it will be found that the colour of the electric light (which can hardly be called a spark, for it fills the tube with light) varies with the kind of gas through which it passes. These exhausted glass tubes are called *vacuum tubes;* but if the air be *entirely* removed from them, it is found that the electric spark cannot pass.

89. The Spiral Tube.—A pretty experiment with the electric spark is made by sticking a number of little diamond-shaped pieces of tinfoil so as to form a spiral within a glass tube (Fig. 58), leaving a space of about one-tenth of an inch between the ends of the pieces of tinfoil. A knob of metal is then fastened to each end of the tube. Holding the tube by one knob, and presenting the other to a charged prime conductor, sparks pass simultaneously between the bits of tinfoil, lighting up the entire tube, and presenting a brilliant appearance when the experiment is performed in a darkened room.

FIG. 58.—Spiral Spangled Tube.

90. Chemical Effects.—(1.) When an electric machine is worked vigorously in dry air, a peculiar smell may be noticed in its neighbourhood. This is owing to the conversion of a part of the ordinary oxygen (which has no smell) of the air into a condensed form of oxygen, called *ozone* (which has a characteristic smell).

(2.) When an electric spark is passed through a mixture of the two gases oxygen and hydrogen, or even through a mixture of coal-gas and air, contained in the *electrical pistol* shown in Fig. 59, an explosion is produced which drives out the cork inserted to keep in the gases.

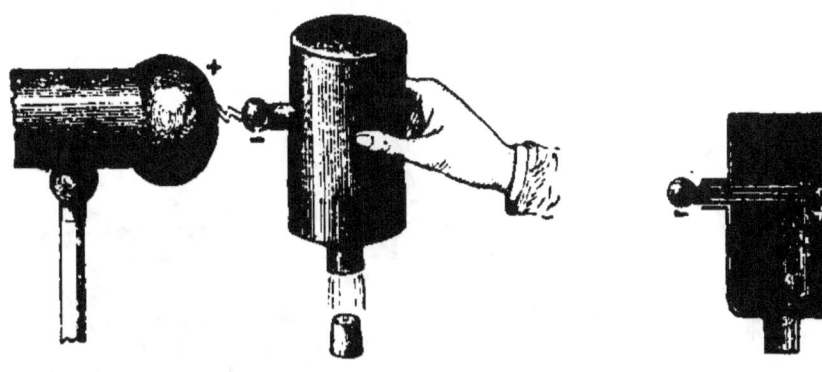

Fig. 59.—Electrical Pistol.   Section of Pistol.

91. Heat Effects.—(1.) When a powerful discharge —as that of a battery of Leyden jars—is made to pass along a thin wire, the heat produced is usually sufficient to melt the wire. Indeed it is found that all conductors are more or less heated by the passage of electricity.

(2.) A jet of coal-gas may be readily lighted by a spark from an electrophorus; and if a little of the inflammable liquid called carbon bisulphide is placed in a warm metal spoon, it will be found easy to ignite it by a single spark.

(3.) To explode gunpowder by the heat of the electric spark, the piece of apparatus known as Henley's universal discharger is commonly employed (Fig. 60). It consists of a little, insulated vulcanite table, a, on each side of which stand the jointed brass rods, c and d, each supported by a pillar of glass, vulcanite, or some other non-conducting material. The gunpowder must be placed upon the table, between the ends of the brass rods, one of which, c, must then be connected with the outer coating of a charged Leyden jar, while the other rod or handle, *d*, is connected by means of the discharging tongs with the knob of the same jar. But the

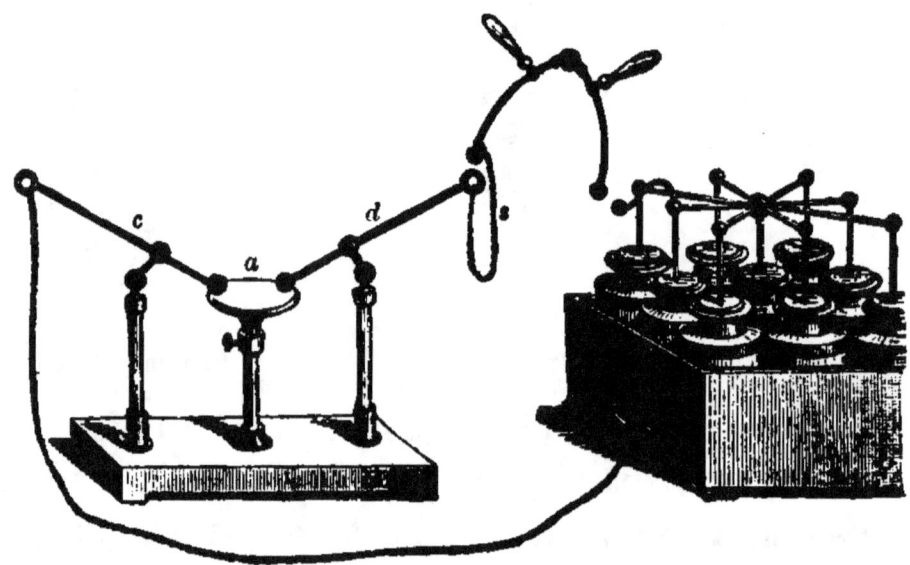

Fig. 60.—Henley's Discharger.

speed at which the electricity then passes through the powder is so great that there is not *time* for the gunpowder to become hot enough to explode. Now the rate at which electricity travels depends partly upon the conducting power of the substance along which it passes. When a weak discharge takes place through a thick copper wire, the velocity is probably as great as that of light, which we know to be 186,000 miles per second. But if we introduce a comparatively bad conductor, such as the piece of wetted string shown at *s* in Fig. 60, we shall *delay* the passage of the electric fluid, and enough heat will then be produced during its passage through the gunpowder to render certain the ignition of the latter substance.

92. Mechanical Effects.—(1.) If a card is placed

between two metal points between which sparks are passing, the card will be perforated. With the

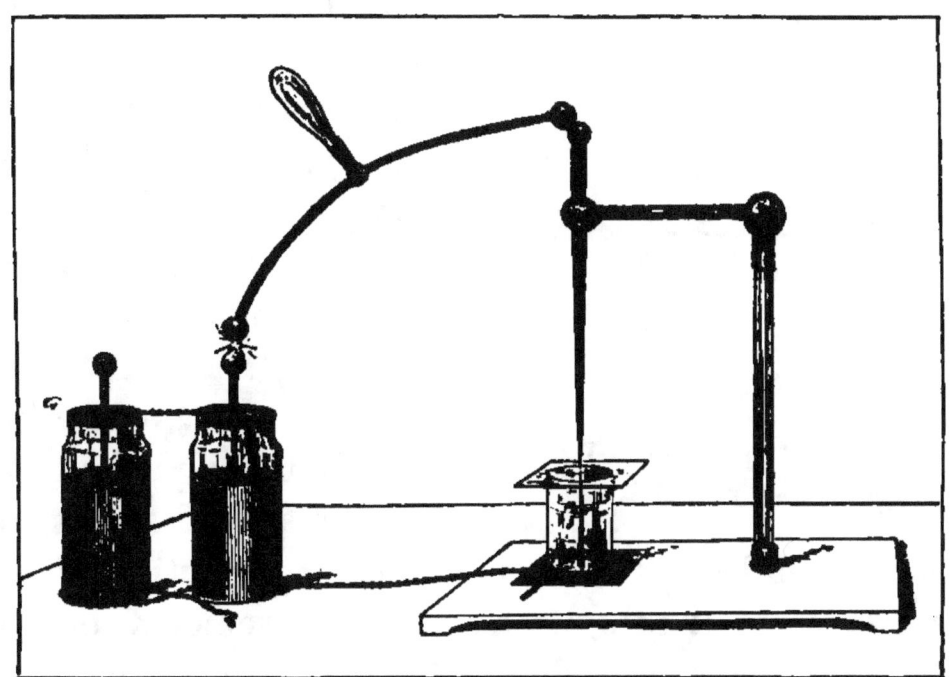

FIG. 61.—Pane of glass pierced by electric spark.

spark from a battery of Leyden jars a sheet of glass can be pierced (Fig. 61) or a piece of dry wood split. (2.) If a spark is passed through a glass vessel filled with water (Fig. 62), the vessel will be shattered by the sudden expansion of the liquid.

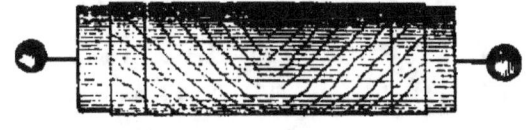

FIG. 62.—Glass tube full of water shattered by spark.

93. Magnetic Effect.—If a common sewing-needle is laid across a strip of tinfoil, and sparks from the prime conductor are then sent along the foil, the needle will become a magnet (Fig. 63). A single discharge of a Leyden battery through the foil will produce the same effect. After the needle has been completely magnetized, it is found that the passage

84 EFFECTS OF ELECTRICITY.

of an electric current in the *opposite* direction will destroy its magnetism. In this way we can understand how it is that the needle of a mariner's compass has been known to have its polarity destroyed or reversed when the ship has been struck by lightning.

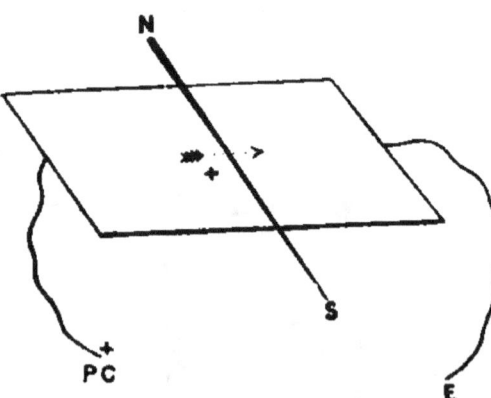

Fig. 63.—Needle laid on strip of tinfoil and magnetized by spark from prime conductor.

94. Physiological Effects.—The passage of electricity through the body of any animal affects the *nerves*, and these in turn produce contraction of the *muscles*, so causing convulsive movements of the limbs. The sparks drawn from a prime conductor produce a pricking sensation in the skin, while the more powerful discharge of a Leyden jar causes a very disagreeable *shock* to be felt in the arms. Several great experimenters have at one time or other received accidentally the discharge of a large battery of Leyden jars through their bodies, the result being that they were rendered senseless for a short time, although they state that they experienced no feeling of pain. It is therefore certain that death by a stroke of lightning must be quite painless. The fact is, the person injured dies before he has *time* to feel any pain.

XV.—ATMOSPHERIC ELECTRICITY.

95. Electricity of the Air—96. Causes of Atmospheric Electricity—97. Discovery of the Cause of Thunderstorms—98. Electricity drawn from the Clouds by Franklin—99. Cause of Lightning and Thunder—100. Distance of Thunderstorms—101. Lightning Conductors—102. Construction of Lightning Conductors—103. Lightning Conductors provide an Easy Path for the Passage of Atmospheric Electricity—104. Silent Discharge from Lightning Conductors—105. The Aurora.

95. Electricity of the Air.—The surface of the Earth, whether land or water, is found usually to have a weak charge of *negative* electricity (see Fig. 64). The layer of air (as a in figure) touching the ground and reaching to about five feet above the surface shows no signs of electricity; but the atmosphere at a greater height than five or ten feet almost invariably shows signs of *positive* electricity (as at b), the amount increasing with the height, and being greatest in clear and dry weather.

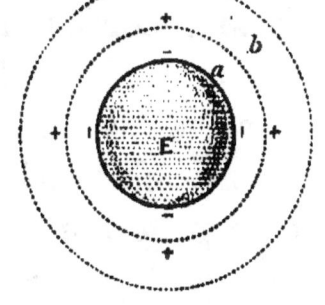

FIG. 64.—The Earth a great Leyden jar.

When we rise to a height of several miles above the Earth's surface, the air is so thin or rare as to be a conductor, like the rarefied air of a vacuum tube. A famous electrician, Lord Kelvin, has well compared the Earth to a great spherical Leyden jar, the upper air and the surface of the ground representing the two coatings, while the layer of dense air next to the ground plays the part of a dielectric (Fig. 64).

96. Causes of Atmospheric Electricity.—(1.) *Evaporation.* Careful experiments show that electricity is produced by the evaporation of liquids. Now the heat of the sun daily causes the evaporation of an enormous quantity of water from the surfaces of the ocean, seas, lakes, rivers, etc. This water-vapour passes into the upper regions of the atmosphere, carrying with it a feeble charge of positive electricity, and leaving the surface of the Earth charged negatively.

(2.) *Friction of Air-Currents.* But although, during fine weather, the upper air is found, almost without exception, to be electrified positively, yet there are many observations showing one portion of the atmosphere to be electrified negatively while another portion is in a positive condition. During storms, certain clouds have been noted to contain negative, while clouds not far distant contained positive, electricity. We can understand that by the friction of currents of unequally heated air against one another, or of moist air against dry air, the two different states of electricity must be produced. When a cold glass rod is rubbed upon a hot surface of glass, the former is found to be +ly and the latter -ly electrified. In the same way the cold winds which blow from the Poles to the Equator may receive a strong positive charge by the friction of this polar air against the currents of hotter air which proceed in a contrary direction. When from this or any other cause the clouds become strongly charged with electricity, they give rise to the phenomena of thunder and lightning.

97. Discovery of the Cause of Thunderstorms.—In the year 1708, an English experimenter, Dr. Wall, published an account of the flashes of light accompanied by a crackling sound, which he obtained by rubbing a large piece of amber with wool. "This light and crackling," says Dr. Wall, "seem in some degree to represent thunder and lightning." In 1729, Stephen Gray also observed the electric spark and the snapping noise which accompanies it. He remarks that "if we are permitted to compare great things with small, these seem to be of the same nature with thunder and lightning."

98. Electricity drawn from the Clouds by Franklin.—It was reserved for Benjamin Franklin to prove in the clearest manner that the lightning flash is identical with the electric spark. He first compared the flash with the spark in every way. The lightning flash is of the same shape as a long electric spark; like that spark, lightning strikes pointed objects in preference to others; lightning pursues the path of least resistance, passing along metals in preference to any other substance; it sets fire to substances, melts metals, tears bodies asunder, produces a sulphureous smell, and even strikes men blind. Franklin imitated experimentally all these effects, striking a pigeon blind and killing a hen and turkey by the electrical discharge from a Leyden battery.

The idea now came into Franklin's mind that it might be possible to tap the clouds during a thunderstorm, and lead their electricity down to the ground. In the year 1752, the great American experimenter completely attained his object by flying a kite

Fig. 65.—Franklin flying his Kite.

during a thunderstorm. The kite had a pointed wire attached to it (Fig. 65), and the electricity descended along the wet string to an iron key which was fastened to the lower end of the string. A piece of silk ribbon served to insulate the kite and string, and Franklin stood beneath a shed in order to keep the silk dry. Sparks were obtained from the key, and these were also used to charge a Leyden jar.

99. Cause of Lightning and Thunder.—The lightning flash heats the air along its path, causing a sudden expansion; the subsequent rush of air into the partial vacuum thus formed produces the noise we call *thunder*. Although a flash of lightning is frequently several miles in length, it does not usually last for more than a fraction of a second.

100. Distance of Thunderstorms.—The lightning and the thunder occur at the same moment, but we *see* the lightning before we *hear* the thunder, because light travels so much faster than sound. As the velocity of light is 186,000 miles per second, the light of the flash may be said to come *instantly* to our eyes; while, as the velocity of sound is on an average only 1,120 feet per second, the thunderclap comes after the flash at a greater or less interval, according to the distance of the storm. This distance may be ascertained by multiplying 1,120 feet by the number of seconds which elapse between seeing the flash and hearing the clap, or, more roughly, by allowing a mile for every five seconds.

101. Lightning Conductors.—Having proved that the force of electricity is the cause of lightning and of thunder, Franklin saw that it was possible to prevent most of the damage often done during a thunderstorm by offering a free passage for the electricity of the clouds to descend and combine with the electricity of the opposite kind upon the surface of the Earth. For this purpose he invented lightning conductors, which have ever since proved of the utmost service in protecting public buildings and property, and in saving human lives from the destructive effects of thunderstorms.

102. Construction of Lightning Conductors.—A lightning conductor consists of a pointed metal rod, which should project two or three feet above the highest part of the building which it is intended to protect (see Fig. 66). The lower end of the rod must be thoroughly connected with the ground, either by fastening it to the system of water-pipes or gas-pipes which run underneath the streets of towns, or by laying it in a stream of water or in a bed of charcoal. The upper end may with advantage have several points

FIG. 66.—Lightning Conductor.

affixed to it, and these should be gilt (or made of platinum), in order to prevent them from rusting. Copper is the best material for a lightning conductor, and the rod should be not less than half-an-inch broad by one quarter of an inch thick.

103. Lightning Conductors provide an Easy Path for the Passage of Atmospheric Electricity.—When the lightning strikes an ordinary brick chimney, it scatters the bricks, and may do much damage to the house in forcing its way through the bricks, stones, mortar, and other non-conducting substances which lie between the cloud and the Earth. But if the chimney be provided with a lightning conductor, the electricity glides safely down the metal, and combines, without doing any harm, with the electricity of the opposite kind which has been attracted, by the inductive action of the charged cloud, to that part of the surface of the Earth which lies immediately beneath the cloud. Ships are protected by a copper tape passing down the main-mast and over the side of the vessel to the sheathing of copper which protects the bottom of the ship.

104. Silent Discharge from Lightning Conductors.—But the most important function of a lightning conductor is to *prevent* any dangerous or disruptive discharge of electricity—any lightning flash—from striking the building to which the conductor is attached. This it does by discharging upwards, from its points, the *opposite* kind of electricity to that of the cloud, thus neutralizing the latter. Sometimes, during violent electrical disturbances of the atmosphere on dark nights, all pointed objects, as the

92 ATMOSPHERIC ELECTRICITY.

masts of ships, the points of lightning conductors, etc., are seen to be tipped by a glow of violet light—St. Elmo's fire, the Italian sailors call it (see Fig. 67)—which is really a brush discharge, caused by the upward passage of electricity attracted from the Earth's surface by the highly electrified clouds. If the cloud, to begin with, is positively electrified, then the lightning conductor allows negative electricity to escape from its points, and this negative, carried up by the molecules of air, combines with the positive electricity upon the water particles which form the cloud. Thus the cloud is rendered neutral, and its electricity is then incapable of doing harm, or indeed of making itself evident in any way.

FIG. 67.—St. Elmo's Fire.

105. The Aurora.—The name of *aurora borealis*, or northern lights, is applied to a beautiful glowing arch of light seen occasionally in the northern

ATMOSPHERIC ELECTRICITY. 93

regions of the Earth (see Fig. 68). To a similar phenomenon which occurs in the southern hemisphere the name of *aurora australis*—southern lights—is given. The streamers of pinkish light which constitute the aurora are believed to be due to currents of electricity traversing the rarefied air at heights of from forty to seventy miles.

Fig. 68.—Aurora Borealis.

APPENDIX.

EDUCATION DEPARTMENT.

FRICTIONAL ELECTRICITY.

NEW CODE: Schedule IV.; Specific Subject, No. xii.

SYLLABUS FOR SECOND STAGE.

Attraction of light bodies by rubbed sealing-wax and glass.

Experimental proof that there are two forms of electricity. Attraction and repulsion.

Gold-leaf electroscope. Construction of electrophorus, electrical machine, and Leyden jar.

Explanation of atmospheric electricity.

EXAMINATION QUESTIONS SET BY H.M. INSPECTORS OF SCHOOLS.

I.

1. Let a pith-ball be suspended by a silk thread to a glass support: on allowing a glass rod rubbed with silk to touch the pith-ball, what happens immediately after contact? If now a stick of sealing-wax rubbed with flannel be brought near the pith-ball, what takes place? Give reasons for your answer; and explain why silk is used for the suspension of the pith-ball.

2. Describe the construction of a glass-plate electrical machine; and show how the prime conductor becomes "charged."

3. If an insulated hollow brass ball be brought near an electrical machine in action, the spark is very feeble; but if the ball be connected with the ground by means of a conductor, the spark will be much more intense. Why?

4. Explain the uses of the following:—
    Proof-plane.
    Discharging tongs.
    Condenser.

APPENDIX. 95

II.

1. You have a glass rod given you. How would you electrify it? and how would you tell which kind of electricity it was charged with?
2. Describe the gold-leaf electroscope.
3. Can you charge an iron rod with electricity? If so, how?
4. How is it that in damp weather it is so difficult to succeed with experiments in frictional electricity?

III.

1. I wish to charge an electroscope with positive electricity. How can I do it by means of sealing-wax rubbed with flannel?
2. I charge an insulated pith-ball with electricity. On bringing my hand near, the ball is attracted. Why?
3. Describe a frictional electrical machine.
4. How would you construct an *electrophorus*? Describe any experiment which you could perform with it.

APPARATUS REQUIRED FOR EXPERIMENTS IN FRICTIONAL ELECTRICITY.

Item	£	s	d	Item	£	s	d
Piece of Amber	0	1	0	Biot's Hemispheres	1	5	0
Two Sticks Sealing-wax	0	2	0	Two Brass Cylinders (insulated)	0	7	6
Amalgam (1 oz.)	0	0	6	Electric Mill	0	3	6
Two Silk Rubbers	0	3	0	Cylinder Machine	1	15	0
Two Flannel Rubbers	0	2	0	Plate Machine (12-in. plate)	2	15	0
Glass Tube (18 in. by 1 in.)	0	1	3	Head of Hair	0	5	0
Vulcanite Rod	0	1	3	Electrical Hail	0	5	6
Pith-balls (1 doz.)	0	0	8	Quadrant Electroscope	0	3	6
Pith-ball Stand	0	3	0	Bevis's Plate	0	3	6
Hank of Silk	0	0	6	Discharging Tongs	0	6	0
Balanced Glass Tube	0	1	3	Epinus's Condenser	0	17	6
Brass Tube with insulating Handle	0	2	6	Open Leyden Jar	0	3	0
Fine Copper Wire	0	0	3	Closed Leyden Jar	0	3	0
Cat's Skin (prepared)	0	4	0	Jar with movable Coatings	0	6	0
Varnished Tumblers (four)	0	2	6	Battery of Leyden Jars (four)	1	0	0
Reel of Silk	0	0	6	Vacuum Tube (egg-shaped)	0	7	0
Gold-leaf Electroscope	0	7	0	Spangled Tube	0	3	6
Proof-plane	0	1	0	Electrical Pistol	0	4	6
Electrophorus (10 in.)	0	10	6	Henley's Discharger	1	2	6
Hollow Brass Ball on insulating Stand	0	3	6	Thin Platinum Wire	0	0	6
Faraday's Net	0	3	6		£14	8	8

The above set can be obtained (or any article separately) from Messrs. SOUTHALL BROS. and BARCLAY, Bull Street, Birmingham, at the prices quoted.

www.ingramcontent.com/pod-product-compliance
Lightning Source LLC
Chambersburg PA
CBHW062357220526
45472CB00008B/1836